Earth at Night, City Lights

This image of Earth at night uses a collection of satellite-based
observations, stitched together in a seamless mosaic of our planet.

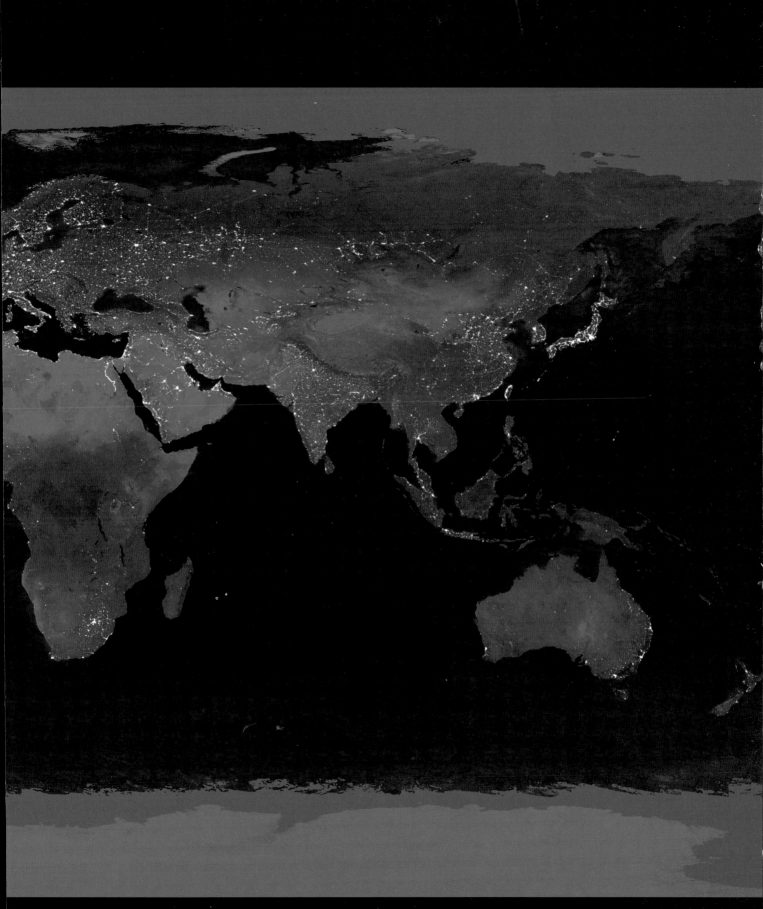

Contemporary Human Geography 2e

Contemporary Human Geography 2e

James M. Rubenstein
MIAMI UNIVERSITY, OXFORD, OHIO

PEARSON

Boston Columbus Indianapolis New York San Francisco Upper Saddle River
Amsterdam Cape Town Dubai London Madrid Milan Munich Paris Montréal Toronto
Delhi Mexico City São Paulo Sydney Hong Kong Seoul Singapore Taipei Tokyo

Pearson
Geography Editor: Christian Botting
Marketing Manager: Maureen McLaughlin
Project Editor: Anton Yakovlev
VP/Executive Director, Development: Carol Trueheart
Development Editors: Jonathan Cheney and Melissa Parkin
Media Producer: Ziki Dekel
Assistant Editor: Kristen Sanchez
Editorial Assistant: Bethany Sexton
Marketing Assistant: Nicola Houston
Managing Editor, Geosciences and Chemistry: Gina M. Cheselka
Project Manager, Production: Maureen Pancza
Project Manager, Full Service: Cindy Miller
Compositor: Element Thomson North America
Senior Technical Art Specialist: Connie Long
Image Lead: Maya Melenchuk
Photo Researcher: Stefanie Ramsay, Bill Smith Group
Operations Specialist: Michael Penne
Cover Photo: Front: Chicago Millennium Park. Back: Aerial view of Chicago.

Dorling Kindersley
Design Development, Page Design, and Layout: Stuart Jackman
Page Layout: Anthony Limerick
Cover Design: Stuart Jackman

Credits and acknowledgments borrowed from other sources and reproduced, with permission, in this textbook appear on the appropriate page within the text.

CIP data available upon request.

6 7 8 9 10—V011—15 14

ISBN-10: 0-321-81112-7; ISBN-13: 978-0-321-81112-7 [Student Edition]
ISBN-10: 0-321-81976-4; ISBN-13: 978-0-321-81976-5 [Instructor's Review Copy]

Brief Contents

Contents

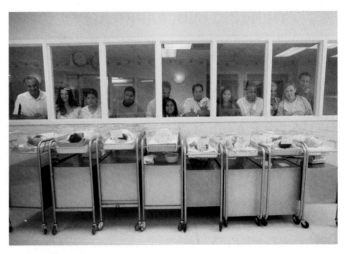

3 MIGRATION

4 FOLK AND POPULAR CULTURE

5 **LANGUAGE**

6 **RELIGION**

7 ETHNICITY

8 POLITICAL GEOGRAPHY

11 INDUSTRY

12 SERVICES AND SETTLEMENTS

13 **URBAN PATTERNS**

14 **RESOURCE ISSUES**

Preface

Welcome to a new kind of geography textbook! We live in a visual age, and geography is a highly visual discipline, so Pearson—the world's leading publisher of geography textbooks—invites you to study human geography as a visual subject.

The second edition of *Contemporary Human Geography* builds on the strengths of the first edition, while responding to user feedback to make important changes and improvements, and incorporating innovative new features, current data, and new information.

NEW TO THIS EDITION

- **Quick Response (QR) Codes.** Each chapter opens with an introductory module that includes a QR code, enabling students on the go to link smartphones from the book to various websites relevant to each chapter, providing easy and immediate access to up-to-date information, data, and statistics from sources such as the United Nations or the U.S. Geological Survey.
- **Key Questions and Main Points.** Each chapter module is framed by conceptual "Key Questions," which ask students to take a bigger picture approach to the concept, and by two "Main Points," which students should understand after studying the module. These Key Questions and Main Points serve as an outline for the topics covered in the chapter and are revisited in the end-of-chapter Review.
- **New and Revised Cartography.** All maps have been thoroughly updated and optimized for maximum accuracy and clear presentation of current data. New projections are used with fewer distortions.
- **Integration of Photos and Text.** The best possible images have been carefully chosen to complement content and concepts. The second edition features well over 400 new photos.

- **The latest science, statistics, and associated imagery.** Data sources include the 2010 U.S. Census and the 2011 Population Reference Bureau World Population Data. Also covers recent political conflicts, economic difficulties, and cultural phenomena such as Facebook and social networking.
- **MapMaster™ Interactive Maps.** These layered thematic interactive maps act as a mini-GIS, allowing students to layer different data at global and regional scales to examine the resulting spatial patterns and practice critical thinking. The interactive maps, with associated tasks and questions, are integrated into select chapter modules and into all end-of-chapter review modules, encouraging students to log in to MasteringGeography to access these exciting geospatial media to practice visual analysis and critical thinking.
- **Google Earth™ Explorations.** Images integrated into select chapter modules and at the end of the chapter pose questions to be answered through Google Earth, the leader in desktop geospatial imagery.
- **Thinking Geographically.** These critical-thinking questions are found at the end of each chapter, giving students a chance to practice higher-order thinking.
- **Looking Ahead.** Each chapter concludes with a brief preview of the next chapter and highlights connections between chapters.
- The new **MasteringGeography™** platform is linked to the Key Questions and Main Points and contains a wide range of assignable and self study resources and activities designed to reinforce basic concepts in human geography, including MapMaster interactive maps, Google Earth activities, geography videos, and more. www.MasteringGeography.com

HUMAN GEOGRAPHY IS CONTEMPORARY

The main purpose of this book is to introduce you to the study of geography as a social science by emphasizing the relevance of geographic concepts to human problems. It is intended for use in college-level introductory human or cultural geography courses. The book is written for students who have not previously taken a college-level geography course.

A central theme in this book is a tension between two important realities of the twenty-first-century world—globalization and cultural diversity. In many respects we are living in a more unified world economically, culturally, and environmentally. The actions of a particular corporation or country affect people around the world. This book argues that, after a period when globalization of the economy and culture has been a paramount concern in geographic analysis, local diversity now demands equal time. People are taking deliberate steps to retain distinctive cultural identities. They are preserving little-used languages, fighting fiercely to protect their religions, and carving out distinctive economic roles.

Recent world events lend a sense of urgency to geographic inquiry. More than a decade into the twenty-first century, we continue to face wars in unfamiliar places and experience economic struggles unprecedented in the lifetimes of students or teachers. Geography's spatial perspectives help to relate economic change to the distributions of cultural features such as languages and religions, demographic patterns such as population growth and migration, and natural resources such as energy, water quality, and food supply.

For example, geographers examine the prospects for an energy crisis by relating the distributions of energy sources and consumption. Geographers find that the users of energy are located in places with different social, economic, and political institutions than the producers of energy. Geographers seek first to describe the distribution of features such as the production and consumption of energy, and then to explain the relationships between these distributions and other human and physical phenomena.

CHAPTER ORGANIZATION

Each chapter is organized into between 9 and 12 two-page modular "spreads" that follow a consistent pattern:

- **Introductory module.** The first spread includes a short introduction to the chapter, as well as an outline of between 9 and 12 issues that will be addressed in the chapter. The key issues are grouped into several overarching Key Questions for that chapter.
- **Topic modules.** Between 9 and 12 modules cover the principal topics of the chapter. Each of these two-page spreads is self-contained and organized around the Key Questions and Main Points, making it easier for an instructor to shuffle the order of presentation. A numbering system also facilitates finding material on a particular spread.
- **Chapter Review modules.** Following the topic modules are concluding spreads that review the chapter's main concepts and key terms while providing students with opportunities to interact with media and engage in critical thinking. The Chapter Review features include:

  o **Key Questions.** The Key Questions presented on the introductory spread are repeated, along with an outline summary of Main Points made in the chapter that address the questions.

  o **On the Internet.** URLs are listed for several useful Internet sites related to the themes of the chapter.

  o **Thinking Geographically.** A thought-provoking idea is introduced, based on concepts and themes developed in the chapter, along with "essay-style" questions.

  o **Interactive Mapping.** Using Pearson's MapMaster interactive mapping media, students create maps and answer questions about spatial relationships of different data. Teachers have the option of assigning these questions in MasteringGeography.

  o **Explore.** Using Google Earth, students inspect imagery from places around the

Innovative End-of-Chapter Tools

Extend student learning with a rich suite of critical-thinking
and media-rich activities.

Review of the Key Questions and the associated Main Points from all modules in the chapter

Thinking Geographically critical-thinking questions

MapMaster™ Interactive Mapping activities

Key Terms definitions

For many ethnicities, sharing space with other ethnicities is difficult, if not impossible. Grievances real and imagined, extending back hundreds of years, prevent peaceful coexistence. Even in countries like the United States, where ethnic diversity is a central feature of the shaping of the American nationality, discriminatory practices cast a long shadow over American history.

Key Questions

Where are ethnicities and races distributed?

► Ethnicity is identity with a group of people who share the cultural traditions of a particular homeland or hearth.

► Race is identity with a group of people who share a biological ancestor.

► Major ethnicities in the United States include African Americans, Hispanic Americans, Asian Americans, and Native Americans.

► Ethnic groups are clustered in regions of the country and within urban neighborhoods.

► African Americans have a distinctive history of forced migration for slavery.

Where are ethnicities and nationalities distributed?

► Nationality is identity with a group of people who share legal attachment and personal allegiance to a particular country.

► A nationality combines an ethnic group's language, religion, and artistic expressions with a country's particular independence movement, history, and patriotism.

Where do ethnicities face conflicts?

► The territory of a nationality rarely matches that inhabited by only one ethnicity.

► Ethnicities compete in many places to dominate territory and control the defining of a nationality.

► Ethnic cleansing is a process in which a more powerful ethnic group forcibly removes a less powerful one in order to create an ethnically homogeneous region.

► Genocide is the mass killing of a group of people in an attempt to eliminate the entire group out of existence.

On the Internet

The U.S. Bureau of the Census provides the most detailed information on the distribution by race and ethnicity in the United States at **www.census.gov**, or scan the QR at the beginning of the chapter.

Thinking Geographically

The U.S. Census permits people to identify themselves as being of more than one race, in recognition that several million American children have parents of two races.

1. What are the merits and difficulties of permitting people to choose more than one race.

Sarajevo, capital of Bosnia & Herzegovina, once contained concentrations of many ethnic groups. In retaliation for ethnic cleansing by the Serbs and Croats, the Bosnian Muslims now in control of Sarajevo have been forcing other ethnic groups to leave the city, and Sarajevo is now inhabited overwhelmingly by Bosnian Muslims (Figure 7.CR.1).

2. What are the challenges in restoring Sarajevo as a city that multiple ethnicities could inhabit?

A century ago European immigrants to the United States had much stronger ethnic ties than today, including clustering in specific neighborhoods..

3. What is the rationale for retaining strong ethnic identity in the United States as opposed to full assimilation into the American nationality?

▼ 7.CR.1 **SARAJEVO WAR GRAVES**

Interactive Mapping

ETHNICITIES AND NATIONALITIES IN SOUTHEAST ASIA

Matching the territory of a nationality to a single ethnicity is rare in the world.

Launch Mapmaster Southeast Asia in Mastering**GEOGRAPHY**

Select: *Cultural* then *Religions.* Select *Cultural* then *Languages* Select *Political* then *Countries*

Can you find an example of a nationality that almost entire encompasses a single ethnicity?

MapMaster™

Explore

MUSEUMS IN DETROIT

Use Google Earth to explore major museums in Detroit that represent ethnic traditions.

Fly to: *Charles H. Wright Museum of African American History, Detroit*

Click 3D Buildings

Drag to: *Enter Street View* in front of the museum.

Exit *Ground Level View* and zoom out until the street and buildings one block to the west are visible.

Click on the Wright museum to see a description of its collection.

Click on the large building in the front left to see what's inside.

How would you compare the collections of the two museums?

Key Terms

Apartheid
Laws (no longer in effect) in South Africa that physically separated different races into different geographic areas.

Balkanization
Process by which a state breaks down through conflicts among its ethnicities.

Centripetal force
An attitude that tends to unify people and enhance support for a state.

Ethnic cleansing
Process by which a more powerful ethnic group forcibly removes a less powerful one in order to create an ethnically homogeneous region.

Ethnicity
Identity with a group of people who share the cultural traditions of a particular homeland or hearth.

Genocide
The mass killing of a group of people in an attempt to eliminate the entire group from existence.

Nationalism
Loyalty and devotion to a particular nationality.

Nationality
Identity with a group of people who share legal attachment and personal allegiance to a particular place as a result of being born there.

Race
Identity with a group of people who share a biological ancestor.

Racism
Belief that race is the primary determinant of human traits and capacities and that racial differences produce an inherent superiority of a particular race.

Racist
A person who subscribes to the beliefs of racism.

Sharecropper
A person who works fields rented from a landowner and pays the rent and repays loans by turning over to the landowner a share of the crops.

Triangular slave trade
A practice, primarily during the eighteenth century, in which European ships transported slaves from Africa to Caribbean islands, molasses from the Caribbean to Europe, and trade goods from Europe to Africa.

► **LOOKING AHEAD**
Ethnicities aspire to political control over areas of Earth through the creation of nation-states, discussed in the next chapter.

On the Internet web links

Google Earth™ Explore image analysis questions

Looking Ahead chapter preview section

Current Data and Applications

The latest science, statistics, and imagery are used in the **Second Edition** for the most contemporary introduction to human geography.

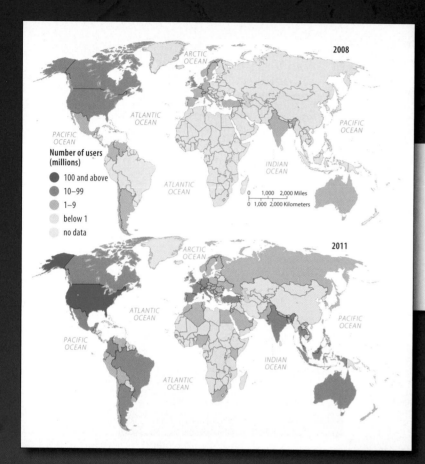

Current data incorporates the latest science, statistics, and imagery, including data from the 2010 U.S. Census, the 2011 World Population Reference Bureau World Population Data, as well as coverage of events like the 2011 "Arab Spring", the birth of South Sudan, and much more. These figures illustrate the explosive growth of Facebook.

Quick Response (QR) Codes on the chapter-opening pages enable students to use smartphones and other mobile devices to link from the book to various open source websites relevant to each chapter, providing easy and immediate access to original sources and up-to-date data.

Students simply scan the QR codes in the book with their mobile smartphones, after following a few easy steps:

1   Download a QR reader from an app store (there are many free apps available) or use the built-in code reader if the device has one.

2   Open the QR code reader app on the phone and scan the code (as shown on the left).

3   The student's device will be automatically redirected to the website. (*Note: data usage charges may apply.*)

Learning Outcomes

MasteringGeography tracks student performance against each instructor's learning outcomes. Instructors can:

- Add their own or use the publisher-provided learning outcomes to track student performance and report it to their administrations.
- View class performance against the specified learning outcomes.
- Export results to a spreadsheet that can be customized further and/or shared with the chair, dean, administrator, or accreditation board.

Mastering offers a data-supported measure to quantify students' learning gains and to share those results quickly and easily.

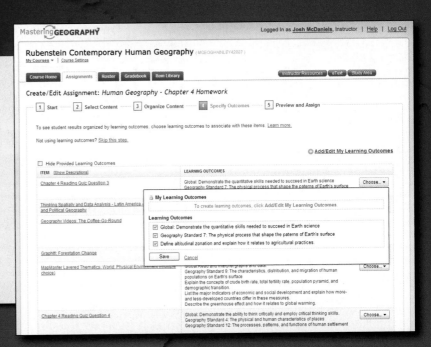

Pearson eText

Pearson eText gives students access to the text whenever and wherever they can access the Internet. The eText pages look exactly like the printed text, and include powerful interactive and customization functions. Users can create notes, highlight text in different colors, create bookmarks, zoom, click hyperlinked words and phrases to view definitions, and view as a single page or as two pages. Pearson eText also links students to associated media files, enabling them to access the media as they read the text, and offers a full text search and the ability to save and export notes.

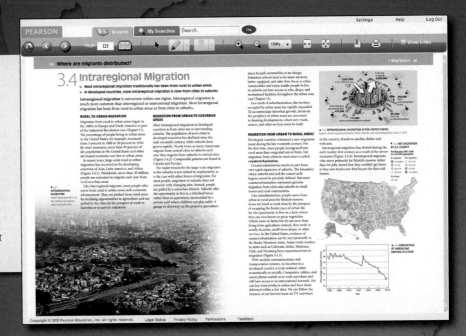

Instructor Resources in MasteringGeography

Assignable activities include:

- MapMaster™ interactive maps
- Geography videos
- Encounter Google Earth™ Explorations
- Reading quizzes
- Test Bank questions
- Thinking Spatially and Data Analysis activities
- End-of-chapter questions and more

1 Thinking Geographically

Thinking geographically is one of the oldest human activities. Perhaps the first geographer was a prehistoric human who crossed a river or climbed a hill, observed what was on the other side, returned home to tell about it, and scratched a map of the route in the dirt. Perhaps the second geographer was a friend or relative who followed the dirt map to reach the other side.

Today, geographers are still trying to understand more about the world in which we live. Geography is the study of where natural environments and human activities are found on Earth's surface and the reasons for their location. This chapter introduces basic concepts that geographers use to study Earth's people and natural environment.

How do geographers describe where things are?

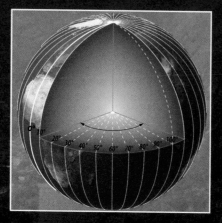

Why is each point on Earth unique?

1.6   **Place: A Unique Location**

1.7   **Region: A Unique Area**

1.1   **Welcome to Geography**

1.2   **Ancient and Medieval Geography**

1.3   **Reading Maps**

1.4   **The Geographic Grid**

1.5   **Geography's Contemporary Analytic Tools**

BEACH AT IPANEMA, BRAZIL

Why does population growth vary among countries?

How might population change in the future?

SCAN FOR UPDATED POPULATION DATA

2.1 Population Concentrations

► **Two-thirds of the world's inhabitants are clustered in four regions.**
► **Humans avoid clustering in harsh environments.**

Human beings are not distributed uniformly across Earth's surface (Figure 2.1.1). Human beings avoid clustering in certain physical environments, especially those that are too dry, too wet, too cold, or too mountainous for activities such as agriculture (Figure 2.1.2).

The clustering of the world's population can be displayed on a cartogram, which depicts the size of countries according to population rather than land area, as is the case with most maps (Figure 2.1.3). Two-thirds of the world's inhabitants are clustered in four regions—East Asia, South Asia, Southeast Asia, and Europe (Figure 2.1.4).

▼ **2.1.1 POPULATION DISTRIBUTION**

Persons per square kilometer

- 1,000 and above
- 250–999
- 25–249
- 5–24
- 1–4
- below 1

▲ **RORAIMA, BRAZIL**

► **2.1.2 SPARSELY POPULATED REGIONS**
Human beings do not live in large numbers in certain physical environments.

■ **2.1.2A COLD LANDS**
Much of the land near the North and South poles is perpetually covered with ice or the ground is permanently frozen (permafrost). The polar regions are unsuitable for planting crops, few animals can survive the extreme cold, and few human beings live there.

■ **2.1.2B DRY LANDS**
Areas too dry for farming cover approximately 20 percent of Earth's land surface. Deserts generally lack sufficient water to grow crops that could feed a large population, although some people survive there by raising animals, such as camels, that are adapted to the climate. Although dry lands are generally inhospitable to intensive agriculture, they may contain natural resources useful to people—notably, much of the world's oil reserves.

■ **2.1.1C WET LANDS**
Lands that receive very high levels of precipitation, such as near Brazil's Amazon River shown in the image, may also be sparsely inhabited. The combination of rain and heat rapidly depletes nutrients from the soil and thus hinders agriculture.

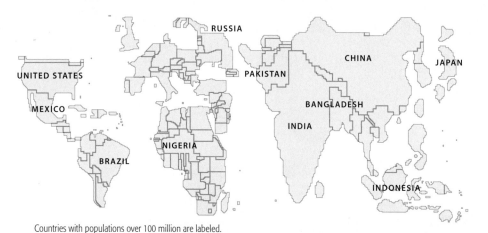

Countries with populations over 100 million are labeled.

▲ **2.1.3 POPULATION CARTOGRAM**
The population cartogram displays the major population clusters of Europe, and East, South, and Southeast Asia as much larger, and Africa and the Western Hemisphere as much smaller than on a more typical equal-area map, such as the large one in the middle of these two pages.

■ **2.1.2D HIGH LANDS**
The highest mountains in the world are steep, snow covered, and sparsely settled. However, some high-altitude plateaus and mountain regions are more densely populated, especially at low latitudes (near the equator) where agriculture is possible at high elevations.

■ 2.1.4A EUROPE

Europe contains one-ninth of the world's people. The region includes four dozen countries, ranging from Monaco, with 1 square kilometer (0.7 square mile) and a population of 32,000, to Russia, the world's largest country in land area when its Asian land portion is included.

Three-fourths of Europe's inhabitants live in cities. A dense network of road and rail lines links settlements. Europe's highest population concentrations are near the major rivers and coalfields of Germany and Belgium, as well as historic capital cities like London and Paris.

The region's temperate climate permits cultivation of a variety of crops, yet Europeans do not produce enough food for themselves. Instead, they import food and other resources from elsewhere in the world. The search for additional resources was a major incentive for Europeans to explore and colonize other parts of the world during the previous six centuries. Today, Europeans turn many of these resources into manufactured products.

▼ 2.1.4 FOUR POPULATION CLUSTERS

The four regions display some similarities. Most of the people in these regions live near an ocean or near a river with easy access to an ocean, rather than in the interior of major landmasses. The four population clusters occupy generally low-lying areas, with fertile soil and temperate climate.

■ 2.1.4B EAST ASIA

One-fifth of the world's people live in East Asia. This concentration includes the world's most populous country, the People's Republic of China. The Chinese population is clustered near the Pacific Coast and in several fertile river valleys that extend inland, such as the Huang and the Yangtze. Much of China's interior is sparsely inhabited mountains and deserts. Although China has 25 urban areas with more than 2 million inhabitants and 61 with more than 1 million, more than one-half of the people live in rural areas where they work as farmers.

In Japan and South Korea, population is not distributed uniformly either. Forty percent of the people live in three large metropolitan areas—Tokyo and Osaka in Japan, and Seoul in South Korea—that cover less than 3 percent of the two countries' land area. In sharp contrast to China, more than three-fourths of all Japanese and Koreans live in urban areas and work at industrial or service jobs.

■ 2.1.4C SOUTHEAST ASIA

A third important Asian population cluster is in Southeast Asia. A half billion people live in Southeast Asia, mostly on a series of islands that lie between the Indian and Pacific oceans. These islands include Java, Sumatra, Borneo, Papua New Guinea, and the Philippines. The largest concentration is on the island of Java, inhabited by more than 100 million people. Indonesia, which consists of 13,677 islands, including Java, is the world's fourth most populous country.

Several islands that belong to the Philippines contain high population concentrations, and people are also clustered along several river valleys and deltas at the southeastern tip of the Asian mainland, known as Indochina. Like China and South Asia, the Southeast Asia concentration is characterized by a high percentage of people working as farmers in rural areas.

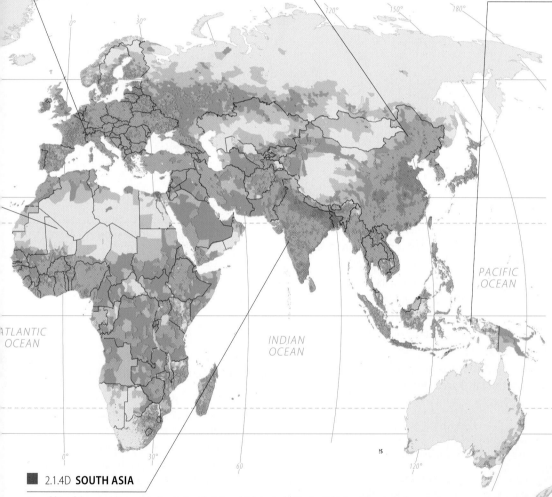

■ 2.1.4D SOUTH ASIA

One-fifth of the world's people live in South Asia, which includes India, Pakistan, Bangladesh, and the island of Sri Lanka. The largest concentration of people within South Asia lives along a 1,500-kilometer (900-mile) corridor from Lahore, Pakistan, through India and Bangladesh to the Bay of Bengal. Much of this area's population is concentrated along the plains of the Indus and Ganges rivers. People are also heavily concentrated near India's two long coastlines—the Arabian Sea to the west and the Bay of Bengal to the east.

To an even greater extent than the Chinese, most people in South Asia are farmers living in rural areas. The region contains 18 urban areas with more than 2 million inhabitants and 46 with more than 1 million, but only one-fourth of the total population lives in an urban area.

▼ VARANASI, INDIA

2.2 Population Density

► **Arithmetic density measures the total number of people living in an area.**

► **Physiological density and agricultural density show spatial relationships between people and resources.**

Density, defined in Chapter 1 as the number of people occupying an area of land, can be computed in several ways, including arithmetic density, physiological density, and agricultural density. These measures of density help geographers to describe the distribution of people in comparison to available resources.

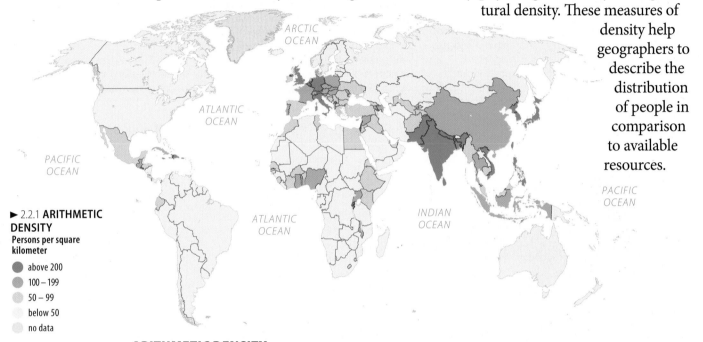

► **2.2.1 ARITHMETIC DENSITY**
Persons per square kilometer

- above 200
- 100 – 199
- 50 – 99
- below 50
- no data

ARITHMETIC DENSITY

Geographers most frequently use **arithmetic density**, which is the total number of people divided by total land area (Figure 2.2.1). Geographers rely on the arithmetic density (also known as *population density*) to compare conditions in different countries because the two pieces of information needed to calculate the measure—total population and total land area—are easy to obtain.

To compute arithmetic density, divide the population by the land area. Figure 2.2.2 shows several examples.

	ARITHMETIC DENSITY (population per square kilometer)	POPULATION 2010 (million people)	LAND AREA (million square kilometers)
Canada	3	34	10.0
United States	32	310	9.6
Netherlands	400	17	0.04
Egypt	80	80	1.0

▲ 2.2.2 **ARITHMETIC DENSITY OF FOUR COUNTRIES**

Compared to the United States, the arithmetic density is much higher in the Netherlands and Egypt and much lower in Canada.

Arithmetic density enables geographers to compare the number of people trying to live on a given piece of land in different regions of the world. Thus, arithmetic density addresses the "where" question. However, to explain why people are not uniformly distributed across Earth's surface, other density measures are more useful (Figure 2.2.3).

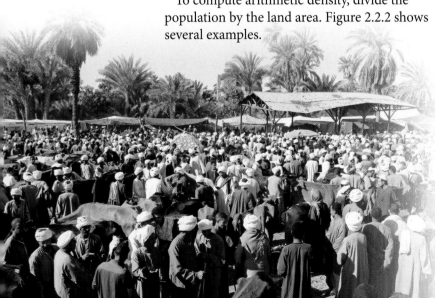

◄ 2.2.3 **HIGH PHYSIOLOGICAL AND AGRICULTURAL DENSITY: EGYPT**
Weekly market at Qutur, Egypt.

PHYSIOLOGICAL DENSITY

A more meaningful population measure is afforded by looking at the number of people per area of **arable land**, which is land suited for agriculture. The number of people supported by a unit area of arable land is called the **physiological density** (Figure 2.2.4). The higher the physiological density, the greater the pressure that people may place on the land to produce enough food.

Physiological density provides insights into the relationship between the size of a population and the availability of resources in a region (Figure 2.2.5). The relatively large physiological densities of Egypt and the Netherlands demonstrate that crops grown on a hectare of land in these two countries must feed far more people than in the United States or Canada, which have much lower physiological densities.

Comparing physiological and arithmetic densities helps geographers to understand the capacity of the land to yield enough food for the needs of the people. In Egypt, for example, the large difference between the physiological density and arithmetic density indicates that most

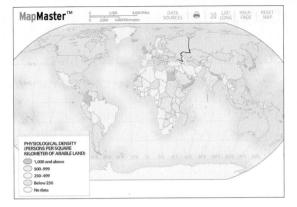

◀ 2.2.4 PHYSIOLOGICAL DENSITY

Open MapMaster World in Mastering**GEOGRAPHY**

Select: *Population* then *Physiological Density.*

What countries other than Egypt and the Netherlands have very high physiological densities?

of the country's land is unsuitable for intensive agriculture. In fact, all but 5 percent of Egyptians live in the Nile River valley and delta, because it is the only area in the country that receives enough moisture (by irrigation from the river) to allow intensive cultivation of crops.

	PHYSIOLOGICAL DENSITY (population per square kilometer of arable land)	ARABLE LAND (million square kilometers)
Canada	65	0.5
United States	175	1.7
Netherlands	1,748	0.01
Egypt	2,296	0.03

◀ 2.2.5 PHYSIOLOGICAL DENSITY OF FOUR COUNTRIES

AGRICULTURAL DENSITY

Two countries can have similar physiological densities, but they may produce significantly different amounts of food because of different economic conditions. **Agricultural density** is the ratio of the number of farmers to the amount of arable land (Figure 2.2.6).

Measuring agricultural density helps account for economic differences. Egypt has a much higher agricultural density than do Canada, the United States, and the Netherlands (Figure 2.2.7). Developed countries have lower agricultural densities because technology and finance allow a few people to farm extensive land areas and feed many people. This frees most of the population in developed countries to work in factories, offices, or shops rather than in the fields.

To understand relationships between population and resources in a country, geographers examine a country's physiological and agricultural densities together. For example, the physiological densities of both Egypt and the Netherlands are high, but the Dutch have a much lower agricultural density than the Egyptians. Geographers conclude that both the Dutch and Egyptians put heavy pressure on the land to produce food, but the more efficient Dutch agricultural system requires fewer farmers than does the Egyptian system.

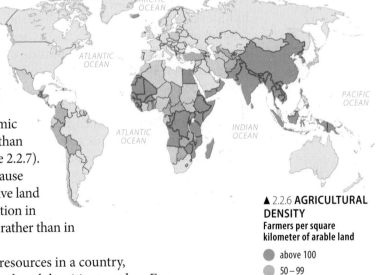

▲ 2.2.6 AGRICULTURAL DENSITY
Farmers per square kilometer of arable land

- above 100
- 50 – 99
- 25 – 49
- below 25
- no data

	AGRICULTURAL DENSITY (farmers per square kilometer of arable land)	PERCENT FARMERS
Canada	1	2
United States	2	2
Netherlands	23	3
Egypt	251	31

▲ 2.2.7 AGRICULTURAL DENSITY OF FOUR COUNTRIES

2.3 Components of Change

► Geographers most frequently measure population change through three indicators.
► Indicators of population change vary widely among regions.

Population increases rapidly in places where many more people are born than die, increases slowly in places where the number of births exceeds the number of deaths by only a small margin, and declines in places where deaths outnumber births. Geographers measure population change in a country or the world as a whole through three measures—crude birth rate, crude death rate, and natural increase rate.

The population of a place also increases when people move in and decreases when people move out. This element of population change—migration—is discussed the next chapter.

NATURAL INCREASE RATE

The **natural increase rate (NIR)** is the percentage by which a population grows in a year. The term natural means that a country's growth rate excludes migration. The world NIR during the early twenty-first century has been 1.2, meaning that the population of the world has been growing each year by 1.2 percent.

About 82 million people are being added to the population of the world annually. That number represents a slight decline from the historic high of 87 million in 1989. The world NIR, though, is considerably lower today than its historic peak of 2.2 percent in 1963. The number of people added each year has declined much more slowly than the NIR because the population base is much higher now than in the past. World population reached 1 billion around 1800. The time needed to add each additional billion has declined (Figure 2.3.1).

The rate of natural increase affects the **doubling time**, which is the number of years

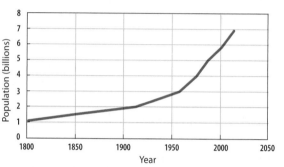

▲ 2.3.1 **WORLD POPULATION GROWTH**

needed to double a population, assuming a constant rate of natural increase. At the early twenty-first-century NIR rate of 1.2 percent per year, world population would double in about 54 years. Should the same NIR continue through the twenty-first century, global population in the year 2100 would reach 24 billion. Should the NIR immediately decline to 1.0, doubling time would stretch out to 70 years, and world population in 2100 would be only 15 billion.

More than 97 percent of the natural increase is clustered in developing countries (Figure 2.3.2). The NIR exceeds 2.0 percent in most countries of sub-Saharan Africa and Southwest Asia & North Africa, whereas it is negative in Europe, meaning that in the absence of immigrants, population actually is declining. About one-third of the world's population growth during the past decade has been in South Asia, one-fourth in sub-Saharan Africa, and the remainder divided about equally among East Asia, Southeast Asia, Latin America, Southwest Asia & North Africa. Regional differences in NIRs show that most of the world's additional people live in the countries that are least able to maintain them.

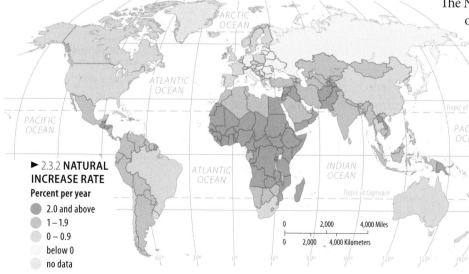

► 2.3.2 **NATURAL INCREASE RATE**
Percent per year
- 2.0 and above
- 1 – 1.9
- 0 – 0.9
- below 0
- no data

Why do people migrate?

What obstacles do immigrants face?

SCAN HERE FOR
U.S. AND WORLD
MIGRATION DATA

3.1 Global Migration Patterns

▶ **Most international migration is from developing countries to developed countries.**

▶ **The United States is the leading destination for international migrants.**

Migration is a permanent move to a new location. It is a form of relocation diffusion, which was defined in Chapter 1 as the spread of a characteristic through the bodily movement of people from one place to another.

Emigration is migration *from* a location; **immigration** (or in-migration) is migration *to* a location. The difference between the number of immigrants and the number of emigrants is the **net migration.**

Geography has no comprehensive theory of migration, although a nineteenth-century outline of 11 migration "laws" written by E. G. Ravenstein is the basis for contemporary geographic migration studies. To understand where and why migration occurs, Ravenstein's "laws" can be organized into three groups: the distance that migrants typically move, the reasons why they move, and their characteristics.

Ravenstein made two main points about the distance that migrants travel to their new homes:

- Most migrants relocate a short distance and remain within the same country (see sections 3.3 and 3.4).

- Long-distance migrants to other countries head for major centers of economic activity (Figure 3.1.1).

▼ 3.1.1 **MIGRATION FROM ASIA TO NORTH AMERICA**
Chinese men waiting outside an employment office in Chinatown, New York, where immigrants seek employment.

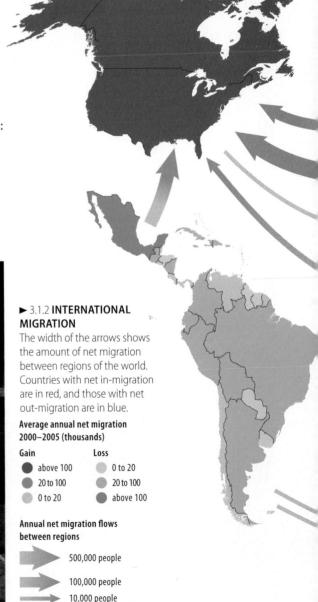

▶ 3.1.2 **INTERNATIONAL MIGRATION**
The width of the arrows shows the amount of net migration between regions of the world. Countries with net in-migration are in red, and those with net out-migration are in blue.

Average annual net migration 2000–2005 (thousands)

Gain		Loss	
⬤	above 100	⬤	0 to 20
⬤	20 to 100	⬤	20 to 100
⬤	0 to 20	⬤	above 100

Annual net migration flows between regions

➡ 500,000 people

➡ 100,000 people

➡ 10,000 people

In the twenty-first century, interregional migration has slowed in the United States (Figure 3.3.2). The severe recession of the early twenty-first century has reduced prospects of finding jobs in other regions of the country.

▶ 3.3.2 **U.S. INTERREGIONAL MIGRATION**
Figures show average annual migration in thousands. The maps show a decrease in total interregional migration in the United States.

1995

500
250
800
750
Northeast
400
300
Midwest
1,600
600
West
1,000 1,600 1,000
1,500
South

2010

90
40
106
259
Northeast
43
53
241 153
Midwest
324 263
West
282
332
South

INTERREGIONAL MIGRATION IN OTHER COUNTRIES

Long-distance interregional migration has opened new regions for development in large countries other than the United States.

- **China.** An estimated 100 million people have migrated from rural areas in the interior of the country to urban areas along the east coast, where jobs are most plentiful, especially in factories. The government once severely limited the ability of Chinese people to make interregional moves, but restrictions have been lifted in recent years (Figure 3.3.3).

- **Russia.** Migration has been encouraged to remote resource-rich regions in Asia through construction of mines, steel mills, power plants, and other industrial enterprises. When controlled by the former Soviet Union, some of the migration was forced.

- **Brazil.** Migration has been encouraged from the large cities along the Atlantic coast to the sparsely settled tropical interior. In 1960, Brazil's capital was moved from the coastal city of Rio de Janeiro to Brasilia, a newly constructed city in the interior (Figure 3.3.4).

- **Indonesia.** Since 1969, the Indonesian government has paid for the migration of more than 5 million people, primarily from the island of Java, where nearly two-thirds of its people live, to less populated islands.

- **India.** A number of governments limit the ability of people to migrate from one region to another. For example, to migrate to India's State of Assam, Indians are required to obtain a permit. Outsiders are limited in order to protect the ethnic identity of Assamese.

▲ 3.3.3 **INTERREGIONAL MIGRATION IN CHINA**
People who migrated from the countryside to Bejing for jobs are sleeping in the Bejing train station while waiting for a train.

▼ 3.3.4 **INTERREGIONAL MIGRATION IN BRAZIL**
When Brasilia became Brazil's capital in 1960, high-rise apartment buildings were constructed to house immigrants from other regions.

3.4 Intraregional Migration

► **Most intraregional migration traditionally has been from rural to urban areas.**

► **In developed countries, most intraregional migration is now from cities to suburbs.**

Intraregional migration is movement within one region. Intraregional migration is much more common than interregional or international migration. Most intraregional migration has been from rural to urban areas or from cities to suburbs.

RURAL TO URBAN MIGRATION

Migration from rural to urban areas began in the 1800s in Europe and North America as part of the Industrial Revolution (see Chapter 11). The percentage of people living in urban areas in the United States, for example, increased from 5 percent in 1800 to 50 percent in 1920. By some measures, more than 90 percent of the population in the United States and other developed countries now live in urban areas.

In recent years, large-scale rural to urban migration has occurred in the developing countries of Asia, Latin America, and Africa (Figure 3.4.1). Worldwide, more than 20 million people are estimated to migrate each year from rural to urban areas.

Like interregional migrants, most people who move from rural to urban areas seek economic advancement. They are pushed from rural areas by declining opportunities in agriculture and are pulled to the cities by the prospect of work in factories or in service industries.

MIGRATION FROM URBAN TO SUBURBAN AREAS

Most intraregional migration in developed countries is from cities out to surrounding suburbs. The population of most cities in developed countries has declined since the mid-twentieth century, while suburbs have grown rapidly. Nearly twice as many Americans migrate from central cities to suburbs each year than migrate from suburbs to central cities (Figure 3.4.2). Comparable patterns are found in Canada and Europe.

The major reason for the large-scale migration to the suburbs is not related to employment, as is the case with other forms of migration. For most people, migration to suburbs does not coincide with changing jobs. Instead, people are pulled by a suburban lifestyle. Suburbs offer the opportunity to live in a detached house rather than an apartment, surrounded by a private yard where children can play safely. A garage or driveway on the property guarantees

▼ 3.4.1
INTRAREGIONAL MIGRATION
Rapid urban growth in La Paz, Bolivia, has spread onto mountainsides.

space to park automobiles at no charge. Suburban schools tend to be more modern, better equipped, and safer than those in cities. Automobiles and trains enable people to live in suburbs yet have access to jobs, shops, and recreational facilities throughout the urban area (see Chapter 13).

As a result of suburbanization, the territory occupied by urban areas has rapidly expanded. To accommodate suburban growth, farms on the periphery of urban areas are converted to housing developments, where new roads, sewers, and other services must be built.

MIGRATION FROM URBAN TO RURAL AREAS

Developed countries witnessed a new migration trend during the late twentieth century. For the first time, more people immigrated into rural areas than emigrated out of them. Net migration from urban to rural areas is called **counterurbanization**.

Counterurbanization results in part from very rapid expansion of suburbs. The boundary where suburbs end and the countryside begins cannot be precisely defined. But most counterurbanization represents genuine migration from cities and suburbs to small towns and rural communities.

Like suburbanization, people move from urban to rural areas for lifestyle reasons. Some are lured to rural areas by the prospect of swapping the frantic pace of urban life for the opportunity to live on a farm where they can own horses or grow vegetables. Others move to farms but do not earn their living from agriculture; instead, they work in nearby factories, small-town shops, or other services. In the United States, evidence of counterurbanization can be seen primarily in the Rocky Mountain states. Some rural counties in states such as Colorado, Idaho, Montana, Utah, and Wyoming have experienced net in-migration (Figure 3.4.3).

With modern communications and transportation systems, no location in a developed country is truly isolated, either economically or socially. Computers, tablets, and smart phones enable us to work anywhere and still have access to an international network. We can buy most products online and have them delivered within a few days. We can follow the fortunes of our favorite team on TV anywhere

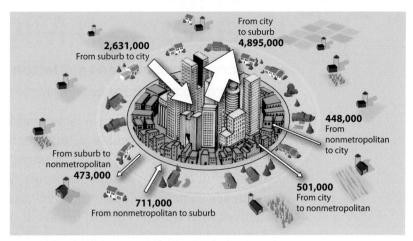

2,631,000 From suburb to city

From city to suburb 4,895,000

448,000 From nonmetropolitan to city

From suburb to nonmetropolitan 473,000

711,000 From nonmetropolitan to suburb

501,000 From city to nonmetropolitan

▲ 3.4.2 **INTRAREGIONAL MIGRATION IN THE UNITED STATES** Figures show migration between cities, suburbs, and nonmetropolitan areas in 2010.

in the country, thanks to satellite dishes and webcasts.

Intraregional migration has slowed during the early twenty-first century as a result of the severe recession (Figure 3.4.4). Intraregional migrants, who move primarily for lifestyle reasons rather than for jobs, found that they couldn't get loans to buy new homes nor find buyers for their old homes.

▼ 3.4.3 **NET MIGRATION BY U.S. COUNTY**

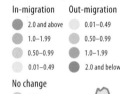

Net Migration 2007–2008 (as % of 2007 population)

In-migration: 2.0 and above; 1.0–1.99; 0.50–0.99; 0.01–0.49
Out-migration: 0.01–0.49; 0.50–0.99; 1.0–1.99; 2.0 and below
No change: 0.00

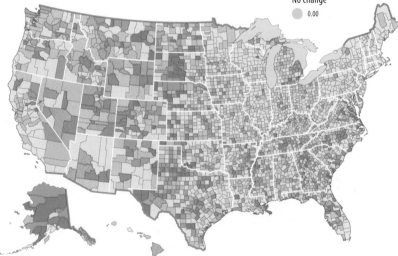

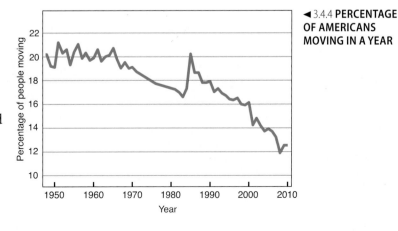

◄ 3.4.4 **PERCENTAGE OF AMERICANS MOVING IN A YEAR**

3.5 Reasons to Migrate

▶ **A combination of push and pull factors influences migration decisions.**

▶ **Most people migrate for economic reasons.**

People migrate because of push factors and pull factors. A **push factor** induces people to move out of their present location, whereas a **pull factor** induces people to move into a new location. As migration for most people is a major step not taken lightly, both push and pull factors typically play a role. To migrate, people view their current place of residence so negatively that they feel pushed away, and they view another place so attractively that they feel pulled toward it. We can identify three major kinds of push and pull factors: political, economic, and environmental.

POLITICAL PUSH AND PULL FACTORS

According to the United Nations, **refugees** are people who have been forced to migrate from their homes and cannot return for fear of persecution because of their race, religion, nationality, membership in a social group, or political opinion. The United Nations High Commissioner for Refugees counted 11 million refugees in 2010 (Figure 3.5.1). Refugees have no home until another country agrees to allow them in, or improving conditions make possible a return to their former home (Figure 3.5.2).

During the eighteenth and early nineteenth centuries, millions of people were shipped to other countries as slaves or as prisoners, especially from Africa to the Western

Hemisphere (see Chapter 7). During the twentieth and twenty-first centuries, wars have forced large-scale migration of ethnic groups, especially in Europe and Africa.

▲ 3.5.2 **POLITICAL MIGRATION: REFUGEES**
Refugees from Sudan in Chad look at a board set up by the Red Cross with pictures of missing children.

▼ 3.5.1 **POLITICAL MIGRATION: REFUGEES**

International refugees
Origin
- ⬤ 1,000,000 and above
- ⬤ 100,000 – 999,999

Destination
- ⬤ 1,000,000 and above
- ⬤ 100,000 – 999,999

Origin and Destination
- ⬤ 100,000 – 999,999

◄ 3.7.3 **EMIGRATION FROM LATIN AMERICA** Children and adults display the Mexican flag at a Latinos Unidos parade in Brooklyn, New York.

Mexican migrants arriving in the United States without proper documents, according to U.S. census and immigration service estimates. But since the 1990s, women have accounted for about half of the undocumented immigrants from Mexico.

AGE AND EDUCATION OF MIGRANTS

The increased female migration to the United States partly reflects the changing role of women in Mexican society: in the past, rural Mexican women were obliged to marry at a young age and to remain in the village to care for children. Now some Mexican women are migrating to the United States to join husbands or brothers already in the United States, but most are seeking jobs. At the same time, women also feel increased pressure to get a job in the United States because of poor economic conditions in Mexico. Immigrants from Latin America and other regions often retain cultural attachment to their home country (Figure 3.7.3). But migrants also adjust to life in America (Figure 3.7.4).

Ravenstein also stated that most long-distance migrants were young adults seeking work, rather than children or elderly people. For the most part, this pattern continues for the United States.

- About 40 percent of immigrants are between the ages of 25 and 39, compared to only 23 percent of the entire U.S. population.

- Only 5 percent of immigrants are over age 65, compared to 12 percent of the entire U.S. population.

- Children under age 15 comprise 16 percent of immigrants, compared to 21 percent for the total U.S. population. However, with the increase in women migrating to the United States, more children are coming with their mothers.

- Recent immigrants to the United States have attended school for fewer years and are less likely to have high school diplomas than are U.S. citizens. The typical undocumented Mexican immigrant has attended school for 4 years, less than the average American but a year more than the average Mexican.

▼ 3.7.4 **FAMILIES OF IMMIGRANTS** Hispanic children at the Wesley Community Center after school program in Phoenix, Arizona.

3.8 Undocumented U.S. Immigrants

▶ **Push and pull factors entice some immigrants to live in the United States without proper authorization.**

▶ **Enforcement of immigration laws is difficult along the U.S.-Mexico border.**

The number of people allowed to immigrate into the United States is at a historically high level, yet the number who wish to come is even higher. Many who cannot legally enter the United States immigrate illegally. Those who do so are entering without proper documents and thus are called **unauthorized (or undocumented) immigrants**. People enter or remain in the United States without authorization primarily because they wish to work but do not have permission to do so from the government.

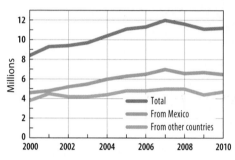

▲ 3.8.1
UNDOCUMENTED IMMIGRANTS IN THE UNITED STATES

The Pew Hispanic Center estimated that there were 11.2 million unauthorized immigrants living in the United States in 2010. The number increased rapidly during the first years of the twenty-first century (Figure 3.8.1). After hitting a peak in 2007, the figure declined because the severe recession starting in 2008 reduced job opportunities in the United States.

Other information about undocumented immigrants, according to Pew Hispanic Center:

• **Source country.** Approximately 60 percent come from Mexico. The remainder are about evenly divided between other Latin American countries and other regions of the world.

• **Children.** The 11.2 million undocumented immigrants included 1 million children. In addition, while living in the United States undocumented immigrants have given birth to approximately 4.5 million babies, who are legal citizens of the United States.

• **Labor force.** Approximately 8 million undocumented immigrants are employed in the United States, accounting for around 5 percent of the total U.S. civilian labor force. Unauthorized immigrants were much more likely than the average American to be employed in construction and hospitality (food service and lodging) jobs and less likely to be in white-collar jobs such as education, health care, and finance.

• **Distribution.** California and Texas have the largest number of undocumented immigrants (Figure 3.8.2). Nevada has the largest percentage.

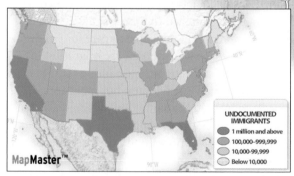

▲ 3.8.2 **UNDOCUMENTED IMMIGRANTS IN THE UNITED STATES**

Individual U.S. states attract immigrants from different countries.

Launch MapMaster North America in Mastering**GEOGRAPHY**

Select: *Political* then *Countries, States, and Provinces*

Select: *Population* then *Destination of Unauthorized Immigrants*. Adjust layer opacity to 50%.

Select: *Population* then *Distribution of African Americans*.

Deselect *African Americans* and select *Hispanic Americans*.

Deselect *Hispanic Americans* and select *Asian Americans*.

Which of the three groups matches most closely with the distribution of states that have the most undocumented immigrants?

POPULAR CULTURE LANDSCAPES: GOLF COURSES

Popular culture can significantly modify or control the environment. It may be imposed on the environment rather than spring forth from it, as with folk culture. In popular culture, the environment may be something to modify for a leisure activity or to promote for the sale of a product. Even if the resulting built environment looks "natural," it may actually be the deliberate creation of people in pursuit of popular social customs.

Golf courses, because of their large size (80 hectares, or 200 acres), provide a prominent example of imposing popular culture on the environment. A surge in popularity spawned construction of roughly 200 golf courses in the United States during the late twentieth century. Geographer John Rooney attributes this to increased income and leisure time, especially among recently retired older people and younger people with flexible working hours.

The provision of golf courses is not uniform across the United States. Although perceived as a warm-weather sport, the number of golf courses per person is actually greatest in the north (Figure 4.8.2). According to Rooney, people in the area have a long tradition of playing golf, and social clubs with golf courses are important institutions in the fabric of the area's popular culture.

The modern game originated as a folk custom in Scotland in the fifteenth century or earlier and diffused to other countries during the nineteenth century. In this respect, the history of golf is not unlike that of soccer described earlier in this chapter. Early Scottish golf courses were primarily laid out on sand dunes adjacent to bodies of water (Figure 4.8.3). Largely because of its origin as a local folk custom, golf courses in Scotland do not modify the environment to the same extent as those constructed in more recent years elsewhere in the world, where hills, sand, and grass are imported, often with little regard for local environmental conditions (Figure 4.8.4).

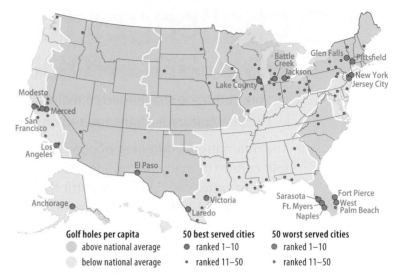

▲ 4.8.2 **POPULAR CULTURE IN THE ENVIRONMENT: GOLF COURSES**

▲ 4.8.3 **FOLK CULTURE OR POPULAR CULTURE? GOLF IN SCOTLAND**

Fly to: *16 Craigend Rd, Troon, Scotland,* to see one of Scotland's most famous golf courses, Royal Troon, founded in 1878.

Drag to enter street view at the square for 16 Craigend Rd.

Click Exit Street View and increase Eye alt to approximately 2,000 ft.

In what ways does this Scottish golf course appear different from typical ones in North America? Does the landscape appear more or less altered at Royal Troon than at golf courses in the United States?

If you would like to look at a U.S. golf course, fly to 8500 River Road, Bethesda, Maryland, to see the one at the Congressional Country Club.

▶ 4.8.4 **GOLF IN THE UNITED STATES**
Westin Mission Hills Resort and Spa in Rancho Mirage near Palm Springs, California.

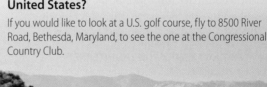

4.9 Challenges to Landscapes of Folk and Popular Culture

▶ Folk culture is challenged to maintain traditional values.

▶ Popular culture is challenged to encourage local diversity.

The international diffusion of popular culture may threaten the survival of traditional folk culture in many places. Many fear the loss of folk culture, especially because rising incomes can fuel demand for the possessions typical of popular culture. At the same time, the uniformity that popular culture has imposed on the landscape is being challenged by the desire of many people to respect and embrace more local diversity.

CHALLENGES TO FOLK CULTURE

The survival of folk culture is threatened in two principal ways:

1. **Loss of traditional values.** When people turn from folk to popular culture, they may also turn away from the society's traditional values. Especially threatened is the subservient role of women to men in some, though not all, folk cultures. Women may have been traditionally relegated to performing household chores, such as cooking and cleaning, and to bearing and raising large numbers of children (Figure 4.9.1).

2. **Imposition of popular culture through diffusion of media.** Exposure to popular culture through the media may stimulate desire to embrace popular culture. Most broadcasting, print, and electronic media emanate from countries where popular culture predominates. Media presents values and behaviors characteristic of popular culture, such as upward social mobility, relative freedom for women, glorification of youth, and stylized violence. A large percentage of the world's countries limit individual freedom to use the Internet, primarily through blocking web and social network sites (Figure 4.9.2).

▼ 4.9.1 **TRADITIONAL FOLK CLOTHING JAPAN**
Most Japanese wear clothing from popular culture, such as jeans and T-shirts. Traditional clothing is reserved mainly for ceremonies and special occasions.

▶ 4.9.2 **LIMITING FREEDOM ON THE INTERNET**
According to OpenNet Initiative, countries limit access to four types of Internet content:

a) Political content that expresses views in opposition to those of the current government, or is related to human rights, freedom of expression, minority rights, and religious movements.

b) Social content related to sexuality, gambling, and illegal drugs and alcohol, as well as other topics that may be socially sensitive or perceived as offensive.

c) Security content related to armed conflicts, border disputes, separatist movements, and militant groups.

d) Internet tools, such as e-mail, Internet hosting, and searching.

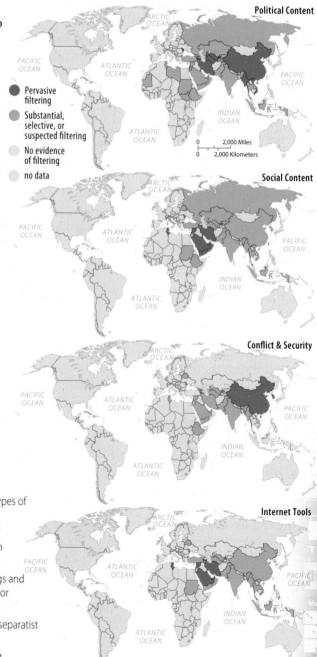

Political Content

● Pervasive filtering

● Substantial, selective, or suspected filtering

No evidence of filtering

no data

0   2,000 Miles
0   2,000 Kilometers

Social Content

Conflict & Security

Internet Tools

▲ 4.9.3 **UNIFORM LANDSCAPE**
A British company emulates American-style fried chicken restaurants, including this one in Swindon, England.

CHALLENGES TO POPULAR CULTURE

The diffusion of popular culture around the world tends to produce more uniform landscapes. The spatial expression of a popular custom in one location will be similar to another. In fact, promoters of popular culture want a uniform appearance to generate product recognition and greater consumption.

Uniformity: Fast Food

The diffusion of fast-food restaurants is a good example of such uniformity. Such restaurants are usually organized as franchises (Figure 4.9.3). The franchise agreement lets the local outlet use the company's name, symbols, trademarks, methods, and architectural styles. To both local residents and travelers, the building and sign are immediately recognizable as part of a national or multinational company.

Uniformity: Gas, Food, and Lodging

Uniformity in the appearance of the landscape is promoted in North America by gas stations, supermarkets, and motels, among other buildings (Figure 4.9.4). These structures are designed so that both local residents and visitors immediately recognize the purpose of the building, even if not the name of the company. American motels and fast-food chains have diffused to other countries. These establishments appeal to North American travelers, yet most customers are local residents who wish to sample American customs they have seen on TV or the Internet.

Diffusion in the Global Marketplace

With faster communications and transportation, customs from any place on Earth can rapidly diffuse elsewhere. Japanese vehicles and electronics, for example, have diffused in recent years to the rest of the world, including North America. Until the 1970s, vehicles produced in North America, Europe, and Japan differed substantially in appearance and size, but in recent years styling has become more uniform, largely because of consumer preference around the world for Japanese vehicles. Carmakers such as General Motors, Ford, Toyota, and Honda now manufacture similar models in North and South America, Europe, and Asia, instead of separately designed models for each continent.

Local Cultures and Globalization

Though globalization has produced a uniform landscape of popular culture, diverse customs survive even in places dominated by popular culture. Even McDonald's, which has come to symbolize uniform popular culture in food service, has increased its recognition of diverse food preferences. The company offers the McMollete in Mexico (English muffin topped with beans, cheese, and salsa) and the McArabia in the Middle East (meat on a pita). In India, where most people are Hindus and avoid consumption of beef, McDonald's serves the McVeggie (rice, bean, and vegetables). The company's "We Buy Local" advertisement campaign has highlighted local sources of produce, such as eggs in Michigan and potatoes in Washington State (Figure 4.9.5).

◀ 4.9.4 **UNIFORM LANDSCAPE**
Billboard jungle along Route 66.

▼ 4.9.5 **ADVERTISING LOCALLY GROWN**
McDonald's, Seattle.

Material culture can be divided into two types—folk and popular. Folk culture is especially interesting to geographers, because it provides a unique identity to each group of people who occupy a specific region of Earth's surface. Popular culture is important, too, because it derives from the high levels of material wealth characteristic of societies that are economically developed. As societies seek to improve their economic level, they may abandon traditional folk culture and embrace popular culture.

Key Questions

How are folk and popular culture distributed?

▶ Differences between folk culture and popular culture can be seen in distinctive forms of leisure.

▶ Folk culture is more likely to have an anonymous origin and to diffuse slowly through migration.

▶ Popular culture is more likely to be invented and diffused rapidly with the use of modern communications.

How are needs of daily life met in folk and popular culture?

▶ Differences between folk and popular culture can be seen in provision of daily necessities, such as food, clothing, and shelter.

▶ Unique regions of folk customs arise because of lack of interaction among groups, even those living nearby.

▶ Popular culture diffuses rapidly; differences in popular culture are more likely to be observed in one place at different points in time than among different places at one point in time.

How is the landscape altered by folk and popular culture?

▶ Rapid diffusion of popular culture has been facilitated by modern communications, especially TV in the twentieth century and the Internet in the twenty-first century.

▶ Folk culture responds to local environmental conditions, whereas popular culture is more likely to alter the landscape.

▶ The diffusion of popular culture can produce uniform landscapes and loss of distinctive local folk customs.

Thinking Geographically

The Amish struggle to maintain their folk culture in the midst of popular culture.

1. Can you give other examples of groups, or isolated individuals, who continue to practice folk culture in a country dominated by popular culture, such as the United States?

Reality TV shows are often set in specific places (4.CR.1).

2. What sorts of folk and popular customs are depicted in reality shows? Are these cultural depictions accurate reflections of the place?

Tourist information is designed to encourage people to visit a particular place.

3. What images of folk and popular culture do countries depict in campaigns to promote tourism? To what extent do these images accurately reflect the countries' culture?

Key Terms

Culture
The body of customary beliefs, social forms, and material traits that together constitute a group of people's distinct tradition.

Custom
The frequent repetition of an act, to the extent that it becomes characteristic of the group of people performing the act.

Folk culture
Culture traditionally practiced by a small, homogeneous, rural group living in relative isolation from other groups.

Habit
A repetitive act performed by a particular individual.

Popular culture
Culture found in a large, heterogeneous society that shares certain habits despite differences in other personal characteristics.

Taboo
A restriction on behavior imposed by social custom.

Terroir
The contribution of a location's distinctive physical features to the way food tastes.

▼ 4.CR.1 *SURVIVOR*: SEASON 3
Shaba National Reserve, Kenya.

EXTINCT LANGUAGES

Thousands of languages are **extinct languages**, once in use—even in the recent past—but no longer spoken or read in daily activities by anyone in the world. *Ethnologue* lists 108 languages that went extinct as recently as the twentieth century and five in the first decade of the twenty-first century. For example, Aka-Bo, a language of the Andamanese family, once spoken in India's Andaman Islands, became extinct in 2010 with the death of its last known speaker.

Hebrew is a rare case of an extinct language that has been revived. A language of daily activity in biblical times, Hebrew diminished in use in the fourth century B.C. and was thereafter retained only for Jewish religious services. At the time of Jesus, people in present-day Israel generally spoke Aramaic, which in turn was replaced by Arabic.

When Israel was established as an independent country in 1948, Hebrew became one of the new country's two official languages, along with Arabic. Hebrew was chosen because the Jewish population of Israel consisted of refugees and migrants from many countries who spoke many languages. Because Hebrew was still used in Jewish prayers, no other language could so symbolically unify the disparate cultural groups in the new country (Figure 5.7.2).

PRESERVING ENDANGERED LANGUAGES

Ethnologue considers approximately 500 languages in danger of becoming extinct, but some are being preserved. The European Union has established the European Bureau for Lesser Used Languages (EBLUL), based in Dublin, Ireland, to provide financial support for the preservation of endangered languages, including several belonging to the Celtic family (Figure 5.7.3):

- **Irish Gaelic.** An official language of the Republic of Ireland, along with English; Ireland's government requires publications be in Irish as well as English.

- **Scottish Gaelic.** Most speakers live in remote highlands and islands of northern Scotland.

- **Welsh.** In Wales, teaching Welsh in schools is compulsory, road signs are bilingual, Welsh-language coins circulate, and a television and radio station broadcast in Welsh.

- **Cornish.** Became extinct in 1777, with the death of the language's last known native speaker; a standard written form of Cornish was established in 2008.

▲ 5.7.2 **REVIVAL OF AN EXTINCT LANGUAGE: HEBREW**
Grocery store, Jerusalem.

- **Breton.** Concentrated in France's Brittany region; Breton differs from the other Celtic languages in that it has more French words.

The survival of any language depends on the political and military strength of its speakers. The Celtic languages declined because the Celts lost most of the territory they once controlled to speakers of other languages.

▼ 5.7.3 **ENDANGERED LANGUAGE: WELSH**
Bilingual parking sign outside The Celtic Manor Resort, Newport, Wales, site of the 2010 Ryder Cup golf tournament.

5.8 French and Spanish in North America

▶ **French and Spanish are increasingly used in North America.**

▶ **Languages can mix to form new ones.**

North America is dominated by English speakers. Yet other languages, especially French in Canada and Spanish in the United States, are becoming increasingly prominent. At the same time, French, Spanish, English, and other languages are mixing to form new languages.

▲ 5.8.1 **FRENCH IN CANADA: "HELLO"**

FRENCH IN CANADA

French is one of Canada's two official languages, along with English (Figure 5.8.1). French speakers comprise one-fourth of the country's population. Most are clustered in Québec, where they comprise more than three-fourths of the province's speakers (Figure 5.8.2).

Until recently, Québec was one of Canada's poorest and least developed provinces. Its economic and political activities were dominated by an English-speaking minority, and the province suffered from cultural isolation and a lack of French-speaking leaders.

The Québec government has made the use of French mandatory in many daily activities. Québec's Commission de Toponyme is renaming towns, rivers, and mountains that have names with English-language origins. The word *Stop* has been replaced by *Arrêt* on the red octagonal road signs, even though *Stop* is used throughout the world, even in France and other French-speaking countries. French must be the predominant language on all commercial signs, and the legislature passed a law banning non-French outdoor signs altogether (later ruled unconstitutional by the Canadian Supreme Court).

Many Québécois favored total separation of the province from Canada as the only way to preserve their cultural heritage. Voters in Québec have thus far rejected separation from Canada, but by a slim majority. Alarmed at these pro-French policies, many English speakers and major corporations moved from Montréal, Québec's largest city, to English-speaking Toronto, Ontario.

Confrontation during the 1970s and 1980s has been replaced in Québec by increased cooperation between French and English speakers. Montréal's neighborhoods, once highly segregated between French-speaking residents on the east and English-speaking residents on the west, have become more linguistically mixed.

Although French dominates over English, Québec faces a fresh challenge of integrating a large number of immigrants from Europe, Asia, and Latin America who don't speak French. Many immigrants would prefer to use English rather than French as their lingua franca but are strongly discouraged from doing so by the Québec government.

Percent French speakers
- 88 – 100
- 63.4 – 87.9
- 25.1 – 63.3
- 5.1 – 25
- 0 – 5
- sparsely inhabited

▲ 5.8.2 **FRENCH/ENGLISH LANGUAGE BOUNDARY IN CANADA**

SPANISH IN THE UNITED STATES

Spanish has become an increasingly important language in the United States because of large-scale immigration from Latin America, as discussed in Chapter 3 (Figure 5.8.3). In some communities, government documents and advertisements are printed in Spanish. Several hundred Spanish-language newspapers and radio and TV stations operate in the United States, especially in southern Florida, the Southwest, and large northern cities (Figure 5.8.4).

Linguistic unity is an apparent feature of the United States, a nation of immigrants who learn English to become U.S. citizens. However, the diversity of languages in the United States is greater than it first appears.

In 2008, a language other than English was spoken at home by 56 million Americans over age 5, 20 percent of the population. Spanish was spoken at home by 35 million people in the United States. More than 2 million spoke Chinese; at least 1 million each spoke French, German, Korean, Tagalog, and Vietnamese. In reaction against the increasing use of Spanish in the United States, 27 states and a number of localities have laws making English the official language.

Americans have debated whether schools should offer bilingual education. Some people want Spanish-speaking children to be educated in Spanish, because they think that children will learn more effectively if taught in their native

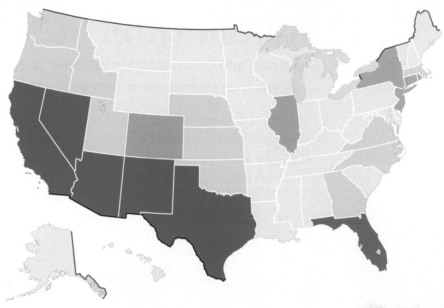

▲ 5.8.3 **DISTRIBUTION OF SPANISH BY U.S. STATES**

Percent of population that speak Spanish at home
- above 15.0
- 9.1 – 15.0
- 5.1 – 9.0
- 1.0 – 5.0

language and that this will also preserve their own cultural heritage. Others argue that learning in Spanish creates a handicap for people in the United States when they look for jobs, virtually all of which require knowledge of English.

Promoting the use of English symbolizes that language is the chief cultural bond in the United States in an otherwise heterogeneous society. With the growing dominance of the English language in the global economy and culture, knowledge of English is important for people around the world, not just inside the United States.

▲ 5.8.4 **SPANISH IN THE UNITED STATES**
Little Havana, Miami, Florida.

CREOLIZED LANGUAGES

A **creole** or **creolized language** is defined as a language that results from the mixing of the colonizer's language with the indigenous language of the people being dominated (Figure 5.8.5). The word *creole* derives from a word in several Romance languages for a slave who is born in the master's house.

A creolized language forms when the colonized group adopts the language of the dominant group but makes some changes, such as simplifying the grammar and adding words from their former language. Creolized language examples include French Creole in Haiti, Papiamento (creolized Spanish) in Netherlands Antilles (West Indies), and Portuguese Creole in the Cape Verde Islands off the African coast.

English has diffused through integration of vocabulary with other languages. The widespread use of English in French is called **Franglais**, in Spanish **Spanglish**, and in German **Denglish**.

▼ 5.8.5 **BISLAMA, A CREOLE LANGUAGE OF VANUATU**
Public health campaign warning sign about AIDS.

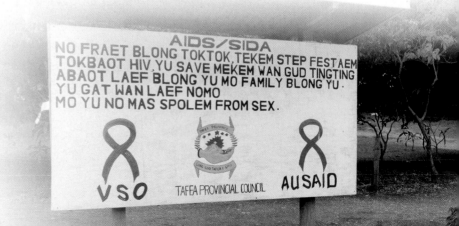

5.9 Multilingual States

▶ **Belgium and Switzerland are examples of multilingual states within Europe.**

▶ **Nigeria is an example of an African country with significant language diversity.**

Difficulties can arise at the boundary between two languages. Note that the boundary between the Romance and Germanic branches of Indo-European runs through the middle of two small European countries, Belgium and Switzerland. Belgium has had more difficulty than Switzerland in reconciling the interests of the different language speakers.

▲ 5.9.1 **LANGUAGE DIVERSITY IN BELGIUM**
Interchange sign in French (first) and Flemish.

BELGIUM

Motorists in Belgium see the language diversity on expressways (Figure 5.9.1). Belgium's language boundary sharply divides the country into two regions. Southern Belgians (known as Walloons) speak French, whereas northern Belgians (known as Flemings) speak a dialect of the Germanic language of Dutch, called Flemish (Figure 5.9.2).

Antagonism between the Flemings and Walloons is aggravated by economic and political differences. Historically, the Walloons dominated Belgium's economy and politics, and French was the official state language. More recently, the Flemings have been better off economically.

In response to pressure from Flemish speakers, Belgium was divided into two independent regions, Flanders and Wallonia. Each elects an assembly that controls cultural affairs, public health, road construction, and urban development in its region.

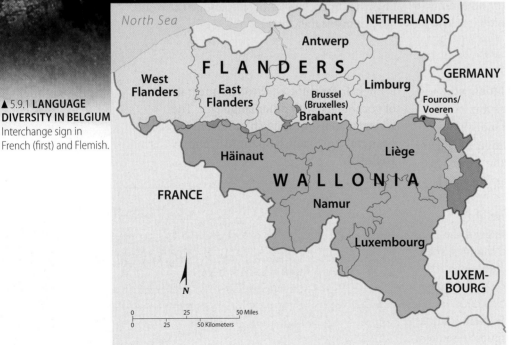

◀ 5.9.2 **LANGUAGES IN BELGIUM**
French is the principal language in Wallonia and Flemish (a dialect of Dutch) in Flanders.

Ethnicities
- Flemings (speaking Dutch dialects)
- Walloons (speaking French)
- Germans
- Flemings and Walloons (legally bilingual)

Protected Minorities
- Walloons in Flanders
- Flemings in Wallonia
- Germans in Wallonia

ISLAM'S JERUSALEM

The most important Muslim structure in Jerusalem is the Dome of the Rock, built in A.D. 691 (Figure 6.10.3). Muslims believe that the large rock beneath the building's dome is the place from which Muhammad ascended to heaven, as well as the altar on which Abraham prepared to sacrifice his son Isaac. Immediately south of the Dome of the Rock is the al-Aqsa Mosque, finished in A.D. 705.

The challenge facing Jews and Muslims is that al-Aqsa Mosque was built on the site of the ruins of the Jewish Second Temple. Israel allows Muslims unlimited access to that religion's holy structures in Jerusalem and some control over them. Through a complex arrangement of ramps and passages patrolled by Palestinian guards, Muslims access the Dome of the Rock and the al-Aqsa Mosque without having to walk in front of the Western Wall where Jews are praying. But with holy Muslim structures sitting literally on top of holy Jewish structures, the two cannot be logically divided by a line on a map (Figure 6.10.4).

▲ 6.10.3 **DOME OF THE ROCK**

▼ 6.10.4 **WESTERN WALL AND DOME OF THE ROCK**
A crowd of Jews are praying at the Western Wall (lower right), situated immediately below the mount containing Islam's Dome of the Rock (top left) and al-Aqsa Mosque (top right).

The Middle East is one of many regions of the world with the potential for conflict resulting from cultural diversity. In the modern world of global economics and culture, the diversity of religions continues to play strong roles in people's lives.

Key Questions
Where are religions distributed?

▶ A religion can be classified as universalizing or ethnic.
▶ Universalizing religions can be divided into branches, denominations, and sects.
▶ Universalizing religions have more widespread distribution than do ethnic religions.

How do religions shape landscapes?

▶ Universalizing religions revere places of importance in the lives of their founders.
▶ Ethnic religions are shaped by the physical geography and agriculture of its hearth.

Where are territorial conflicts between religions?

▶ Long-standing conflicts among religious groups can be found in a number of regions.
▶ Religious conflicts in the Middle East have been especially long-standing and intractable.

▼ 6.CR.1 **PEOPLE PRAYING AT THE FRIDAY PRAYER IN THE DITIB-MERKEZ MOSQUE, DUISBURG-MARXLOH, NORTH RHINE-WESTPHALIA, GERMANY, EUROPE**

Thinking Geographically

Sharp demographic differences, such as NIR, CBR, and net migration, can be seen among Jews, Christians, and Muslims in the Middle East.

1. **How might demographic differences affect future relationships among the religious groups in the region?**

Islam seems strange and threatening to some people in predominantly Christian countries (Figure 6.CR.1).

2. **To what extent is this attitude shaped by knowledge of the teachings of Muhammad and the Quran, and to what extent is it based on lack of knowledge of the religion?**

People carry their religious beliefs with them when they migrate. Over time, change occurs in the regions from which most U.S. immigrants originate and in the U.S. regions where they settle.

3. **How has the distribution of U.S. religious groups been affected by these changes?**

On the Internet

Statistics on the number of adherents to religions, branches, and denominations are at **www.adherents.com** or by scanning the QR on the opening page of this chapter.

Glenmary Research Center, which is affiliated with the Roman Catholic Church, provides maps of U.S. religions at **www.glenmary.org**. Glenmary has a map of Americans not affiliated with any religion (high percentage in red).

Interactive Mapping

FORCED MIGRATION IN SOUTH ASIA

Millions of people were forced to migrate after South Asia gained independence from the United Kingdom in 1947.

Launch Mapmaster South Asia in **Mastering GEOGRAPHY**

Select: *Religions* from the *Cultural* menu, then *Ethnic Division* from the *Population* menu, then *Countries and States* from the *Political* menu.

What accounts for the migration pattern?

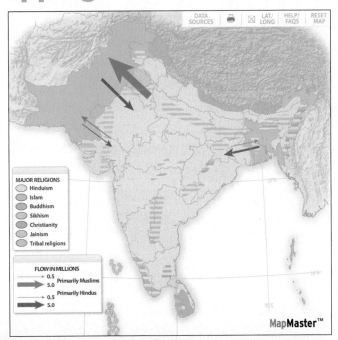

DATA SOURCES | LAT/ LONG | HELP/ FAQS | RESET MAP

MAJOR RELIGIONS
- Hinduism
- Islam
- Buddhism
- Sikhism
- Christianity
- Jainism
- Tribal religions

FLOW IN MILLIONS
- 0.5
- 5.0 Primarily Muslims
- 0.5
- 5.0 Primarily Hindus

MapMaster™

Imagery Date: 2/18/2010 2004   21°25'14.27" N 39°49'35.00" E elev 308 m   Eye alt 1.35 km

Image © 2011 GeoEye   ©2010 Google

Explore

MAKKAH, SAUDI ARABIA

Use Google Earth to explore Masjid al-Haram, Islam's largest mosque in Makkah, Saudi Arabia. Millions of Muslims make a pilgrimage to Makkah each year and gather at Masjid al-Haram Mosque.

Fly to: *Masjid al-Haram Mosque, Makkah, Saudi Arabia.*

Drag to: *Enter Street View* to the square in the middle of the mosque

Click to look around so that North is at the bottom.

Click 3D buildings.

Continue to look around to see the tall building with the clock tower immediately south of the Mosque.

1. **What is in the tall building?**

2. **Why would this building be located immediately next to the mosque?**

Key Terms

Animism
Belief that objects, such as plants and stones, or natural events, like thunderstorms and earthquakes, have a discrete spirit and conscious life.

Branch (of a religion)
A large and fundamental division within a religion.

Cosmogony
A set of religious beliefs concerning the origin of the universe.

Denomination (of a religion)
A division of a branch that unites a number of local congregations in a single legal and administrative body.

Ethnic religion
A religion with a relatively concentrated spatial distribution whose principles are likely to be based on the physical characteristics of the particular location in which its adherents are concentrated.

Fundamentalism
A literal interpretation and a strict and intense adherence to basic principles of a religion.

Missionary
An individual who helps to diffuse a universalizing religion.

Monotheism
The doctrine or belief of the existence of only one god.

Pilgrimage
A journey to a place considered sacred for religious purposes.

Polytheism
Belief in or worship of more than one god.

Sect (of a religion)
A relatively small group that has broken away from an established denomination.

Universalizing religion
A religion that attempts to appeal to all people, not just those living in a particular location.

▶ **LOOKING AHEAD**

The next chapter continues our look at the world's cultural patterns by examining ethnic diversity at several scales.

EUROPEANS ONLY

7 Ethnicity

Each of us belongs to one or more ethnic groups with which we share important attributes. Ethnicity is an especially important element of culture, because our ethnic identity is irrefutable. We can deny or suppress our ethnicity, but we cannot choose to change it in the same way we can choose to speak a different language or practice a different religion. If our parents come from two ethnic groups or our grandparents from four, our ethnic identity may be extremely diluted, but it never completely disappears.

Ethnic identity is a source of pride to people, a link to the experiences of ancestors and to cultural traditions, such as food and music preferences. Ethnicity also matters in places with a history of discrimination by one ethnic group against another. Even without discriminatory practices, ethnic groups may vary according to life expectancy, infant mortality, and other important measures.

Where are ethnicities and races distributed?

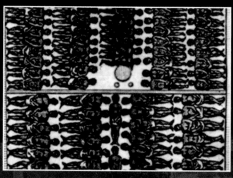

7.1 **Race and Ethnicity**

7.2 **Distribution of Ethnicities in the United States**

7.3 **African American Migration**

7.4 **Discrimination by Race**

APARTHEID MUSEUM,
JOHANNESBURG,
SOUTH AFRICA

EUROP

How are states organized?

How do states interact with each other?

SCAN TO ACCESS CIA DATA ABOUT EVERY COUNTRY

8.1 A World of States

▶ **The world is divided into nearly 200 states.**
▶ **All but a handful of states belong to the United Nations.**

A **state** is an area organized into a political unit and ruled by an established government that has control over its internal and foreign affairs. A state occupies a defined territory on Earth's surface and contains a permanent population. A state has **sovereignty**, which means control of its internal affairs without interference by other states. Because the entire area of a state is managed by its national government, laws, army, and leaders, it is a good example of a formal or uniform region. The term *country* is a synonym for state.

The term *state*, as used in political geography, does not refer to local governments, such as those inside the United States. The 50 "states" that comprise the United States are subdivisions within a single state—the United States of America.

A map of the world shows that nearly all states belong to the United Nations (Figure 8.1.1). When it was founded in 1945, only around 50 states were members of the UN (Figure 8.1.2). Membership grew to 193 states in 2011. The two most populous states not in the United Nations in 2011 were Kosovo (Figure 8.1.3) and Taiwan (Figure 8.1.4).

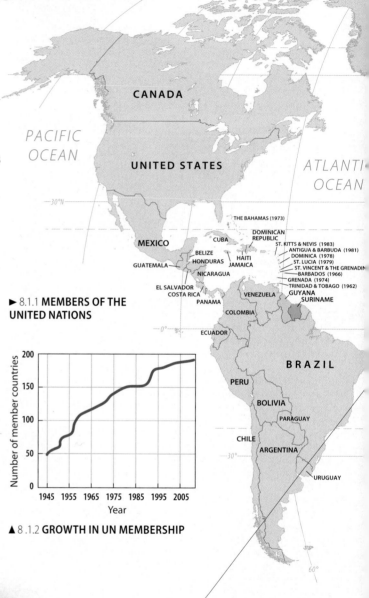

▶ 8.1.1 **MEMBERS OF THE UNITED NATIONS**

▲ 8.1.2 **GROWTH IN UN MEMBERSHIP**

◀ 8.1.3 **KOSOVO: A SOVEREIGN STATE?**
(left) The Republic of Kosovo declared its independence from Serbia in 2008, following ethnic cleansing and war crimes by some Serb leaders (refer to section 7.8). The United States and most European countries recognize Kosovo as an independent sovereign state, but Serbia, Russia, and most countries of Africa and Asia do not. (far left) Pristina, capital of Kosovo.

193 members

- Original members: 51
- 1940s: Added 8
- 1950s: Added 24
- 1960s: Added 42
- 1970s: Added 25
- 1980s: Added 7
- 1990s: Added 31
- 2000s: Added 4
- 2010s: Added 1
- Nonmember

◄ 8.1.4 **TAIWAN: A SOVEREIGN STATE?**

(left) The governments of most other states consider China (officially, the People's Republic of China) and Taiwan (officially, the Republic of China) as separate and sovereign states. According to China's government, Taiwan is not sovereign, but a part of China. This confusing situation arose from a civil war in China during the late 1940s between the Nationalists and the Communists. After losing, nationalist leaders in 1949 fled to Taiwan, 200 kilometers (120 miles) off the Chinese coast.

The Nationalists proclaimed that they were still the legitimate rulers of the entire country of China. Until some future occasion when they could defeat the Communists and recapture all of China, the Nationalists argued, at least they could continue to govern one island of the country. The United Nations transferred China's seat from the Republic of China to the People's Republic of China in 1971 and the United States transferred diplomatic recognition to the People's Republic in 1979. (right) Taipei, capital of Taiwan.

8.2 Ancient States

► **City-states originated in ancient times in the Fertile Crescent.**
► **States developed in Europe through consolidation of kingdoms.**

The concept of dividing the world into a collection of independent states is relatively recent. Prior to the 1800s, Earth's surface was organized in other ways, such as city-states, empires, and tribes. Much of Earth's surface consisted of unorganized territory.

ANCIENT STATES

The development of states can be traced to the ancient Middle East, in an area known as the Fertile Crescent.

The ancient Fertile Crescent formed an arc between the Persian Gulf and the Mediterranean Sea (Figure 8.2.1). The eastern end, Mesopotamia, was centered in the valley formed by the Tigris and Euphrates rivers, in present-day Iraq. The Fertile Crescent then curved westward over the desert, turning southward to encompass the Mediterranean coast through present-day Syria, Lebanon, and Israel. The Nile River valley of Egypt is sometimes regarded as an extension of the Fertile Crescent. Situated at the crossroads of Europe, Asia, and Africa, the Fertile Crescent was a center for land and sea communications in ancient times.

A **city-state** is a sovereign state that comprises a town and the surrounding countryside. Walls clearly delineated the boundaries of the city, and outside the walls the city controlled agricultural land to produce food for urban residents. The countryside also provided the city with an outer

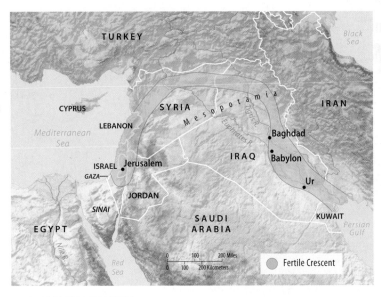

▲ 8.2.1 **FERTILE CRESCENT**

line of defense against attack by other city-states. The first states to evolve in Mesopotamia were known as city-states.

In Mesopotamia, periodically, one city-state would gain military dominance over the others and form an empire. Mesopotamia was organized into a succession of empires by the Sumerians, Assyrians, Babylonians, and Persians.

Meanwhile, the state of Egypt emerged as a separate empire to the west of the Fertile Crescent. Egypt controlled a long, narrow region along the banks of the Nile River, extending from the Nile Delta at the Mediterranean Sea southward for several hundred kilometers. Egypt's empire lasted from approximately 3000 B.C. until the fourth century B.C.

Ancient Greece also consisted of a collection city-states, including Athens, Corinth, and Sparta (Figure 8.2.2). The Greek city-states lost their independence during the fourth century B.C., when they were ruled by Macedonian kings, beginning with Philip II (382–336, ruled 359–336) and his son Alexander III ("Alexander the Great" 356–323, ruled 336–323).

▼ 8.2.2 **ANCIENT ATHENS**
The Parthenon, a temple to the goddess Athena, patron of the Athens city-state, is on the hilltop Acropolis (citadel) to the right. To the left is the Temple of Hephastaeus, the god of technology.

▲ 8.2.3 **ANCIENT ROME**
The most prominent building is the Temple of Jupiter, dedicated to the king of the ancient gods.

EUROPEAN STATES

Political unity in the ancient world reached its height with the establishment of the Roman Empire (Figure 8.2.3). The Roman Empire controlled most of Europe, North Africa, and Southwest Asia, from modern-day Spain to Iran and from Egypt to England (Figure 8.2.4). At its maximum extent, the empire comprised 38 provinces, each using the same set of laws that were created in Rome. Massive walls helped the Roman army defend many of the empire's frontiers (Figure 8.2.5).

The Roman Empire collapsed in the fifth century after a series of attacks by people living on its frontiers and because of internal disputes. The European portion of the Roman Empire was fragmented into a large number of estates owned by competing kings, dukes, barons, and other nobles. Victorious nobles seized control of defeated rivals' estates, and after these nobles died, others fought to take possession of their

▼ 8.2.5 **HADRIAN'S WALL**.
The Roman army built the wall beginning in A.D. 122 in northern Britain, possibly to prevent invasion or to control immigration.

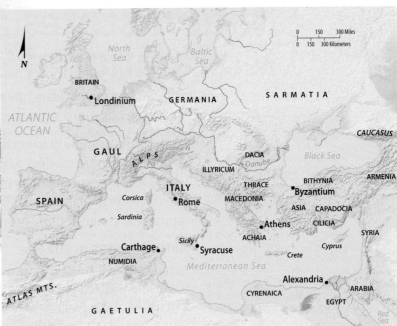

▲ 8.2.4 **ROMAN EMPIRE IN** A.D. **100**

land. Meanwhile, most people were forced to live on an estate, working and fighting for the benefit of the noble.

A handful of powerful kings emerged as rulers over large numbers of estates beginning about the year 1100. The consolidation of neighboring estates under the unified control of a king formed the basis for the development of such modern European states as England, France, and Spain. However, much of central Europe—notably present-day Germany and Italy—remained fragmented into a large number of estates that were not consolidated into states until the nineteenth century (Figure 8.2.6).

▼ 8.2.6 **EUROPE IN 1300**

8.3 Nation-states and Multinational States

► In the modern world, many states have been created to encompass nationalities.
► A multinational state has more than one ethnicity recognized as distinct nationalities.

▲ 8.3.1 **NATION-STATES IN EUROPE**

Many ethnic groups hold strong desires for **self-determination**, which is the right to govern themselves within sovereign states. A **nation-state** is a state whose territory corresponds to that occupied by a particular ethnicity that has been transformed into a nationality. Through the creation of nation-states, the aspiration of many ethnic groups for self-determination has been realized.

NATION-STATES IN EUROPE

Europe was transformed during the nineteenth and twentieth centuries from a fragmented collection of kingdoms, principalities, and empires (Figure 8.3.1, top) into nation-states (Figure 8.3.1, second). Boundaries between states were fixed to conform as closely as possible to those of leading ethnic groups.

Denmark is a fairly good example of a nation-state, because the territory occupied by the Danish ethnicity closely corresponds to the state of Denmark. But even Denmark is not a perfect example of a nation-state. The country's 80-kilometer (50-mile) southern boundary with Germany does not divide Danish and German ethnic groups precisely. To dilute the concept of a nation-state further, Denmark controls two territories in the Atlantic Ocean that do not share Danish cultural characteristics—the Faeroe Islands and Greenland (Figure 8.3.2).

Creating a German nation-state has proved especially challenging. Central Europe was a patchwork of hundreds of small states until the most powerful of them—Prussia—forged a German Empire in 1871. The boundaries of Germany were altered drastically twice during the twentieth century after losses in World War I (1914–18) and World War II (1939–45).

During the 1930s, German National Socialists (Nazis) claimed that all German-speaking parts of Europe constituted one nationality and should be unified into one state. They pursued this goal forcefully, and other European powers did not attempt to stop the Germans from taking over Austria and the German-speaking portion of Czechoslovakia, known as the Sudetenland. Not until the Germans invaded Poland (clearly not a German-speaking country) in 1939 did England and France try to stop them, marking the start of World War II.

After it was defeated in World War II, Germany was divided into two countries (Figure 8.3.1, third). Two Germanys existed from 1949 until 1990. With the collapse of communism in Europe, the German Democratic Republic ceased to exist, and its territory became part of the German Federal Republic (Figure 8.3.1, bottom).

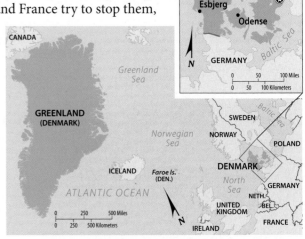

▲ 8.3.2 **DENMARK**

MULTINATIONAL STATES

A **multinational state** contains two or more ethnic groups with traditions of self-determination. Relationships among nationalities vary in different multinational states. In some states, one nationality tries to dominate another, especially if one of the nationalities greatly outnumbers the other, whereas in other states nationalities coexist peacefully.

The Union of Soviet Socialist Republics (U.S.S.R.) was an especially prominent example of a multinational state until its collapse in 1991 (Figure 8.3.3). The Soviet Union's 15 republics were based on its 15 largest ethnicities. The 15 largest ethnicities of the former Soviet Union are now independent countries that represent varying degrees of nation-states:

- **Three Baltic states: Estonia, Latvia, and Lithuania.** These three small neighbors have differences in language and religion and distinct historical traditions. They had been independent until annexed to the Soviet Union in 1940.

- **Three European states: Belarus, Moldova, Ukraine.** Belarusians and Ukrainians became distinctive ethnicities because they were isolated from the main body of Eastern Slavs—the Russians—between the thirteenth and eighteenth centuries. Moldovans are ethnically indistinguishable from Romanians.

- **Five Central Asian states: Kazakhstan, Kyrgyzstan, Tajikistan, Turkmenistan, and Uzbekistan.** The "stans" are predominantly Muslim and speak Altaic languages (except for Tajiks who speak a language similar to Persian).

- **Three Caucasus states: Armenia, Azerbaijan, and Georgia.** Armenians are Eastern Orthodox Christians who speak a separate branch of Indo-European. Azeris (or Azerbaijanis) are Muslims who speak an Altaic language. The two nation-states have clashed over their shared boundary. Georgians are Eastern Orthodox Christians. Two ethnicities within Georgia, the Ossetians and Abkhazianas, are fighting for autonomy and possible reunification with Russia.

- **Russia: Now the world's largest multinational state.** Russia identifies 21 national republics that are supposed to be the homes of its largest ethnicities, but the government recognizes in some way at least 170 ethnicities (Figure 8.3.4). Overall, 20 percent of the country's population is non-Russian. Particularly troublesome for the Russians are the ethnicities bordering the Caucasus states, especially the Chechens and Ossetians.

▲ 8.3.3 **STATES IN THE FORMER U.S.S.R.**

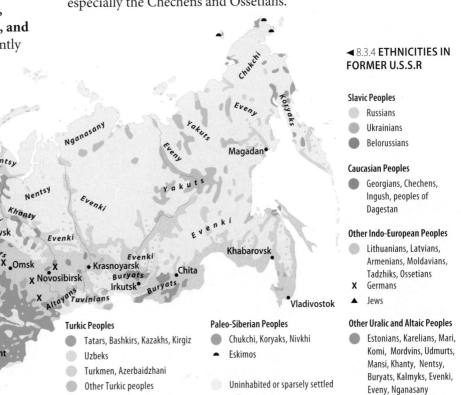

◄ 8.3.4 **ETHNICITIES IN FORMER U.S.S.R**

Slavic Peoples
- Russians
- Ukrainians
- Belorussians

Caucasian Peoples
- Georgians, Chechens, Ingush, peoples of Dagestan

Other Indo-European Peoples
- Lithuanians, Latvians, Armenians, Moldavians, Tadzhiks, Ossetians
- X Germans
- ▲ Jews

Turkic Peoples
- Tatars, Bashkirs, Kazakhs, Kirgiz
- Uzbeks
- Turkmen, Azerbaidzhani
- Other Turkic peoples

Paleo-Siberian Peoples
- Chukchi, Koryaks, Nivkhi
- Eskimos
- Uninhabited or sparsely settled

Other Uralic and Altaic Peoples
- Estonians, Karelians, Mari, Komi, Mordvins, Udmurts, Mansi, Khanty, Nentsy, Buryats, Kalmyks, Evenki, Eveny, Nganasany

8.4 Challenges in Defining States

▶ The sovereignty of some land area is disputed.
▶ International treaties cover possession of polar and coastal regions.

Most of the world has been allocated to sovereign states. A handful of places test the definition of sovereignty.

KOREA: ONE STATE OR TWO?

A colony of Japan for many years, Korea was divided into two occupation zones by the United States and the former Soviet Union after they defeated Japan in World War II (Figure 8.4.1). The division of these zones became permanent in the late 1940s, when the two superpowers established separate governments and withdrew their armies. The new government of the Democratic People's Republic of Korea (North Korea) then invaded the Republic of Korea (South Korea) in 1950, touching off a 3-year war that ended with a cease-fire line near the 38th parallel.

Both Korean governments are committed to reuniting the country into one sovereign state. However, progress toward reconciliation has been hindered by North Korea's decision to build nuclear weapons, even though the country has lacked the ability to provide its citizens with food, electricity, and other basic needs. Meanwhile, in 1992, North Korea and South Korea were admitted to the United Nations as separate countries.

▲ 8.4.1 **NORTH AND SOUTH KOREA**
A nighttime satellite image recorded by the U.S. Air Force Defense Meteorological Satellite Program shows the illumination of electric lights in South Korea, whereas North Korea has virtually no electric lights, a measure of its poverty and limited economic activity.

WESTERN SAHARA (SAHRAWI REPUBLIC)

The Sahrawi Arab Democratic Republic, also known as Western Sahara, is considered by most African countries as a sovereign state. Morocco, however, claims the territory and to prove it has built a 2,700-kilometer sand wall (known as a berm) around the territory to keep out rebels (Figure 8.4.2).

Spain controlled the territory on the continent's west coast between Morocco and Mauritania until withdrawing in 1976. An independent Sahrawi Republic was declared by the Polisario Front and recognized by most African countries, but Morocco and Mauritania annexed the northern and southern portions, respectively. Three years later Mauritania withdrew, and Morocco claimed the entire territory.

Morocco controls most of the populated area, but the Polisario Front operates in the vast, sparsely inhabited deserts, especially the one-fifth of the territory that lies east of Morocco's wall. The United Nations has tried but failed to reach a resolution between the parties.

▼ 8.4.2 **WESTERN SAHARA**
Morocco built a 2,700-kilometer-long wall to bolster its claim on the Western Sahara.

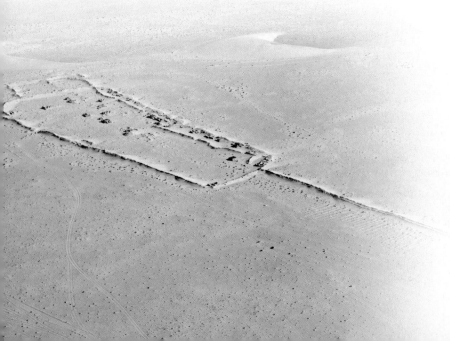

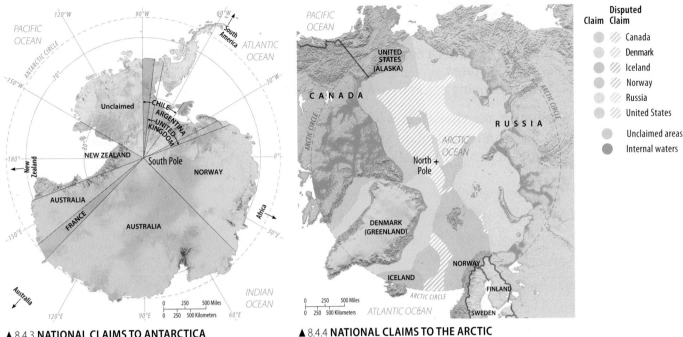

▲ 8.4.3 **NATIONAL CLAIMS TO ANTARCTICA**

▲ 8.4.4 **NATIONAL CLAIMS TO THE ARCTIC**

POLAR REGIONS: MANY CLAIMS

Antarctica is the only large land mass on Earth's surface that is not part of a state. Several states, including Argentina, Australia, Chile, France, New Zealand, Norway, and the United Kingdom, claim portions of Antarctica (Figure 8.4.3). Argentina, Chile, and the United Kingdom have made conflicting, overlapping claims. The United States, Russia, and a number of other states do not recognize the claims of any country to Antarctica. The Antarctic Treaty, signed in 1959, provides a legal framework for managing Antarctica. States may establish research stations there for scientific investigations, but no military activities are permitted. The treaty has been signed by 47 states.

As for the Arctic, the 1982 United Nations Convention on the Law of the Sea permitted countries to submit claims inside the Arctic Circle by 2009 (Figure 8.4.4). The Arctic region is thought to be rich in energy resources.

THE LAW OF THE SEA

The United Nations Convention on the Law of the Sea, signed by 158 countries, has defined waters extending various distances from the coastlines of states (Figure 8.4.5). Disputes can be taken to a Tribunal for the Law of the Sea or to the International Court of Justice.

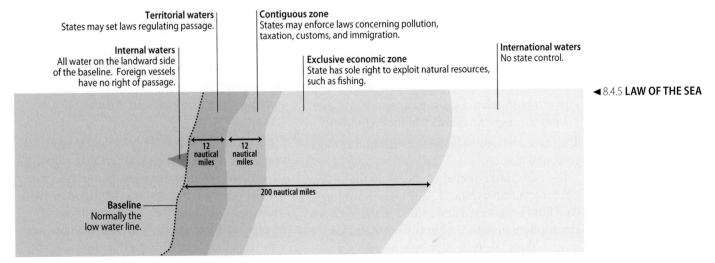

◀ 8.4.5 **LAW OF THE SEA**

8.5 Colonies

▶ **Until the twentieth century, much of the world consisted of colonies of European states.**

▶ **Most remaining colonies are islands with small populations.**

A **colony** is a territory that is legally tied to a sovereign state rather than being completely independent. In some cases, a sovereign state runs only the colony's military and foreign policy. In others, it also controls the colony's internal affairs.

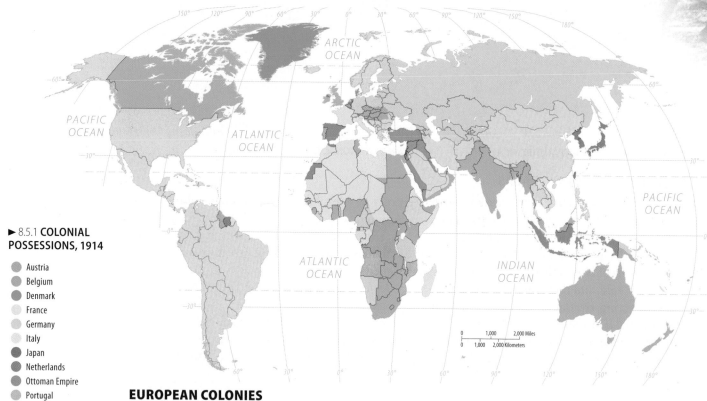

▶ 8.5.1 **COLONIAL POSSESSIONS, 1914**

- Austria
- Belgium
- Denmark
- France
- Germany
- Italy
- Japan
- Netherlands
- Ottoman Empire
- Portugal
- Russia
- Spain
- United Kingdom
- United States

EUROPEAN COLONIES

European states came to control much of the world through **colonialism**, which is the effort by one country to establish settlements in a territory and to impose its political, economic, and cultural principles on that territory (Figure 8.5.1). European states established colonies elsewhere in the world for three basic reasons:

- To promote Christianity.
- To extract useful resources.
- To establish relative prestige among European states through the number of their colonies.

The three motives could be summarized as God, gold, and glory.

The European colonial era began in the 1400s, when European explorers sailed westward for Asia but encountered and settled in the Western Hemisphere instead. The European states lost most of their Western Hemisphere colonies

after independence was declared in 1776 by the United States and by most Latin American states between 1800 and 1824.

European states then turned their attention to Africa and Asia (Figure 8.5.2).

- The United Kingdom established colonies in every region of the world, and proclaimed that "the Sun never sets on the British Empire."

- France had the second-largest overseas territory, primarily in West Africa and Southeast Asia.

Most African and Asian colonies became independent after World War II. Only 15 African and Asian states were members of the United Nations when it was established in 1945, compared to 106 in 2011. The boundaries of the new states frequently coincide with former colonial provinces, although not always.

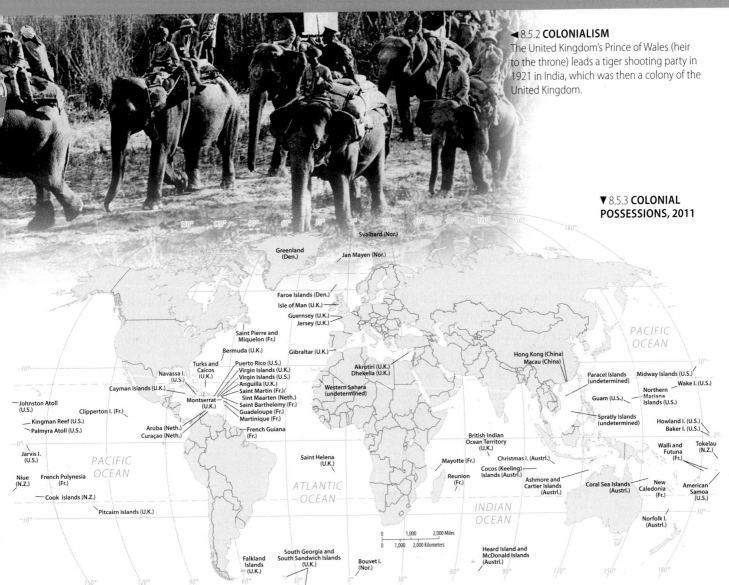

◀ 8.5.2 **COLONIALISM**
The United Kingdom's Prince of Wales (heir to the throne) leads a tiger shooting party in 1921 in India, which was then a colony of the United Kingdom.

▼ 8.5.3 **COLONIAL POSSESSIONS, 2011**

REMAINING COLONIES

The U.S. Department of State lists 68 places in the world that it calls dependencies and areas of special sovereignty (Figure 8.5.3). Most are islands in the Pacific Ocean or Caribbean Sea. The list includes 43 with indigenous populations and 25 with no permanent population.

The State Department list includes several entities that others do not classify as colonies:

- Greenland. Part of Denmark, it has a high degree of autonomy and self-rule and even makes independent foreign policy decisions.

- Hong Kong and Macao (Figure 8.5.4). Special Administrative Regions of China, with autonomy in economic matters but not in foreign and military affairs. Hong Kong was a colony of the United Kingdom until 1997, and Macao of Portugal until 1999.

On the other hand, the State Department list does not include several inhabited islands considered by other sources to be colonies, including Australia's Lord Howe Island, Britain's Ascension Island, and Chile's Easter Island.

▼ 8.5.4 **FLAGS OF HONG KONG**
Special Administrive Region (lower left) and People's Republic of China.

8.6 Shapes of States

▶ States have one of five basic shapes.
▶ States that have no water boundary are landlocked.

The shape of a state controls the length of its boundaries with other states. The shape therefore affects the potential for communication and conflict with neighbors. Countries have five basic shapes—compact, prorupted, elongated, fragmented, and perforated. Examples of each can be seen in sub-Saharan Africa. Each shape displays distinctive characteristics and challenges.

ELONGATED STATES: POTENTIAL ISOLATION

A handful of **elongated states** have a long and narrow shape. Examples in sub-Saharan Africa include :

- Gambia, which extends along the banks of the Gambia River about 500 kilometers (300 miles) east–west but is only about 25 kilometers (15 miles) north–south (Figure 8.6.1).
- Malawi, which measures about 850 kilometers (530 miles) north–south but only 100 kilometers (60 miles) east–west.

Elsewhere in the world, Chile and Italy are prominent examples. Elongated states may suffer from poor internal communications. A region located at an extreme end of the elongation might be isolated from the capital, which is usually placed near the center.

▲ 8.6.1 GAMBIA: AN ELONGATED STATE

FRAGMENTED STATES: PROBLEMATIC

A **fragmented state** includes several discontinuous pieces of territory. Technically, all states that have offshore islands as part of their territory are fragmented. However, fragmentation is particularly significant for some states. There are two kinds of fragmented states:

- Fragmented states separated by an intervening state. An example in sub-Saharan Africa is Angola, which is divided into two fragments by the Congo Democratic Republic. An independence movement is trying to detach Cabinda as a separate state from Angola, with the justification that its population belongs to distinct ethnic groups (Figure 8.6.2).

▲ 8.6.2 ANGOLA: A FRAGMENTED STATE

- Fragmented states separated by water. An example in sub-Saharan Africa is Tanzania, which was created in 1964 as a union of the island of Zanzibar with the mainland territory of Tanganyika (Figure 8.6.3). Although home to different ethnic groups, the two entities agreed to join together because they shared common development goals and political priorities.

Prominent examples of fragmented states elsewhere in the world include Russia (which has a fragment called Kaliningrad) and Indonesia (which comprises 13,677 islands).

▼ 8.6.3 TANZANIA: A FRAGMENTED STATE
Unguja (Zanzibar Island), part of Tanzania, includes Zanzibar City. Stone Town, shown here, is the old part of the city.

PRORUPTED STATES: ACCESS OR DISRUPTION

An otherwise compact state with a large projecting extension is a **prorupted state**. Proruptions are created for two principal reasons:

1. To provide a state with access to a resource, such as water. For example, in southern Africa, Congo has a 500-kilometer (300-mile) proruption to the west along the Zaire (Congo) River. The Belgians created the proruption to give their colony access to the Atlantic (Figure 8.6.4).

2. To separate two states that otherwise would share a boundary. For example, in southern Africa, Namibia has a 500-kilometer (300-mile) proruption to the east called the Caprivi Strip. When Namibia was a colony of Germany, the proruption disrupted communications among the British colonies of southern Africa. It also provided the Germans with access to the Zambezi, one of Africa's most important rivers.

▲ 8.6.4 **CONGO AND NAMIBIA: PRORUPTED STATES**

COMPACT STATES: EFFICIENT

In a **compact state**, the distance from the center to any boundary does not vary significantly. Compactness facilitates establishing good communications to all regions, especially if the capital is located near the center. Examples of compact states in sub-Saharan Africa include Burundi, Kenya, Rwanda, and Uganda (Figure 8.6.5). Compactness does not necessarily mean peacefulness, as compact states are just as likely as others to experience civil wars and ethnic rivalries.

▲ 8.6.5 **SUB-SAHARAN AFRICA: SEVERAL COMPACT STATES**

LANDLOCKED STATES

A **landlocked state** lacks a direct outlet to the sea because it is completely surrounded by other countries (only one country in the case of Lesotho). Landlocked states are especially common in sub-Saharan Africa (Figure 8.6.6). The prevalence of landlocked states in Africa is a remnant of the colonial era, when the United Kingdom and France held much of the region as colonies. As independent countries, landlocked states had to cooperate with neighboring coastal states in order to bring in supplies and ship out minerals. Railroads built by the European powers became critical in connecting landlocked states with seaports in neighboring states.

▲ 8.6.6 **SUB-SAHARAN AFRICA: SEVERAL LANDLOCKED STATES**

PERFORATED STATES: COMPLETELY SURROUNDING

A state that completely surrounds another one is a **perforated state**. For example, South Africa completely surrounds Lesotho (Figure 8.6.7). Lesotho must depend almost entirely on South Africa for the import and export of goods. Dependency on South Africa was especially difficult for Lesotho when South Africa had a government controlled by whites who discriminated against the black majority population. Italy is another prominent example, as it surrounds the Holy See (the Vatican) and San Marino.

▲ 8.6.7 **SOUTH AFRICA: A PERFORATED STATE**

8.7 Boundaries

▶ **Physical boundaries include mountains, deserts, and bodies of water.**
▶ **Cultural boundaries include geometric and ethnic boundaries.**

A state is separated from its neighbors by a **boundary**, an invisible line marking the extent of a state's territory. When looking at satellite images of Earth, we see physical features like mountains and oceans, but not boundaries between countries. Boundary lines are not painted on Earth, but they might as well be, because for many people they are more meaningful than natural features.

Boundaries are of two types:

- Physical boundaries coincide with significant features of the natural landscape.

- Cultural boundaries follow the distribution of cultural characteristics.

Neither type of boundary is better or more "natural," and many boundaries are a combination of both types.

PHYSICAL BOUNDARIES

Important physical features on Earth's surface can make good boundaries because they are easily seen, both on a map and on the ground. Three types of physical elements serve as boundaries between states (Figure 8.7.1):

- **Desert Boundaries.** Deserts make effective boundaries because they are hard to cross and sparsely inhabited. In North Africa, the Sahara has generally proved to be a stable boundary separating Algeria, Libya, and Egypt on the north from Mauritania, Mali, Niger, Chad, and the Sudan on the south.

- **Mountain Boundaries.** Mountains can be effective boundaries if they are difficult to cross. Contact between nationalities living on opposite sides may be limited, or completely impossible if passes are closed by winter storms. Mountains are also useful boundaries because they are rather permanent and are usually sparsely inhabited.

- **Water Boundaries.** Rivers, lakes, and oceans are commonly used as boundaries, because they are readily visible on maps and aerial imagery. Historically, water boundaries offered good protection against attack from another state, because an invading state had to transport its troops by ship and secure a landing spot in the country being attacked. The state being invaded could concentrate its defense at the landing point.

▲▶ 8.7.1 **PHYSICAL BOUNDARIES**
(above) Desert boundary between Libya and Chad. (right) Mountain boundary between Argentina and Chile. (far right) Water boundary between Germany and France.

CULTURAL BOUNDARIES

Two types of cultural boundaries are common—geometric and ethnic. Geometric boundaries are simply straight lines drawn on a map. Ethnic boundaries between states coincide with differences in ethnicity, as well as language and religion.

- **Geometric Boundaries.** Part of the northern U.S. boundary with Canada is a 2,100-kilometer (1,300-mile) straight line (more precisely, an arc) along 49° north latitude, running from Lake of the Woods between Minnesota and Manitoba to the Strait of Georgia between Washington State and British Columbia (Figure 8.7.2). This boundary was established in 1846 by a treaty between the United States and the United Kingdom, which then controlled Canada. The two countries share an additional 1,100-kilometer (700-mile) geometric boundary between Alaska and the Yukon Territory along the north–south arc of 141° west longitude.

- **Ethnic Boundaries.** Boundaries between countries have been placed where possible to separate ethnic groups. Language is also an important cultural characteristic for drawing boundaries, especially in Europe. Religious differences often coincide with boundaries between states, but in only a few cases has religion been used to select the actual boundary line (Figure 8.7.3).

▲ 8.7.2 **GEOMETRIC BOUNDARY: UNITED STATES AND CANADA** International Peace Park between North Dakota and Manitoba.

▲ 8.7.3 **ETHNIC BOUNDARY: GREEK AND TURKISH CYPRUS**
Cyprus, the third-largest island in the Mediterranean Sea, contains two nationalities—Greek and Turkish. Several Greek Cypriot military officers who favored unification of Cyprus with Greece seized control of the government in 1974. Shortly after, Turkey invaded Cyprus to protect the Turkish Cypriot minority, and the portion of the island controlled by Turkey declared itself the independent Turkish Republic of Northern Cyprus in 1983.

FRONTIERS

A **frontier** is a zone where no state exercises complete political control. A frontier is an area often many kilometers wide that is either uninhabited or sparsely settled. Historically, frontiers rather than boundaries separated many states (Figure 8.7.4). Almost universally, frontiers between states have been replaced by boundaries. Modern communications systems permit countries to monitor and guard boundaries effectively, even in previously inaccessible locations. Once-remote frontier regions have become more attractive for agriculture and mining.

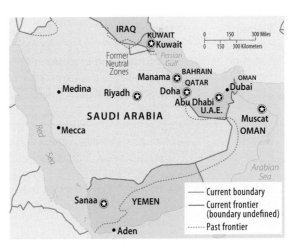

◀ 8.7.4 **FRONTIERS: ARABIAN PENINSULA**

8.8 Governing States

▶ **Some governments are more democratic than others.**

▶ **Two types of government structures are unitary states and federal states.**

A state has two types of government—a national government and local governments. At the national scale, a government can be more or less democratic. At the local scale, the national government can determine how much power to allocate to local governments.

NATIONAL GOVERNMENT REGIMES

National governments can be classified as democratic, autocratic, or anocratic (Figure 8.8.1). An **autocracy** is a country that is run according to the interests of the ruler rather than the people. An **anocracy** is a country that is not fully democratic or fully autocratic, but rather displays a mix of the two types. According to the Center for Systemic Peace, a democracy and an autocracy differ in three essential elements (Figure 8.8.2).

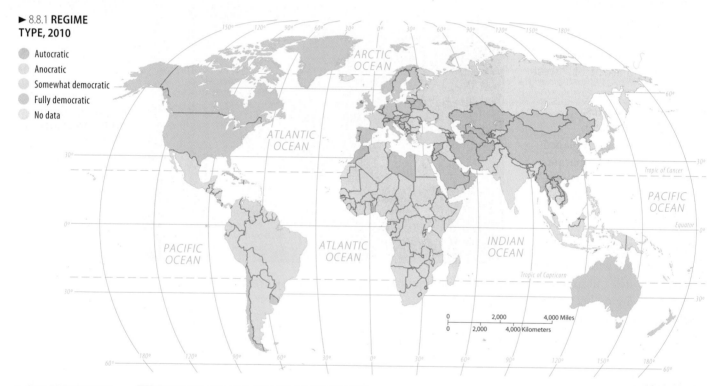

▶ 8.8.1 **REGIME TYPE, 2010**

- Autocratic
- Anocratic
- Somewhat democratic
- Fully democratic
- No data

▶ 8.8.2 **DIFFERENCES BETWEEN DEMOCRACY AND AUTOCRACY**

Element	Democracy	Autocracy
Selection of leaders	Institutions and procedures through which citizens can express effective preferences about alternative policies and leaders.	Leaders are selected according to clearly defined (usually hereditary) rules of succession from within the established political elite.
Citizen participation	Institutionalized constraints on the exercise of power by the executive.	Citizens' participation is sharply restricted or suppressed.
Checks and balances	Guarantee of civil liberties to all citizens in their daily lives and in acts of political participation.	Leaders exercise power with no meaningful checks from legislative, judicial, or civil society institutions.

TREND TOWARDS DEMOCRACY

In general, the world has become more democratic (Figure 8.8.3). The Center for Systemic Peace cites three reasons:

- The replacement of increasingly irrelevant and out-of-touch monarchies with elected governments that are able to regulate, tax, and mobilize citizens in exchange for broadening individual rights and liberties.
- The widening of participation in policy making to all citizens through universal rights to vote and to serve in government (Figure 8.8.4).
- The diffusion of democratic government structures created in Europe and North America to other regions of the world.

▲ 8.8.3 **DEMOCRACY TREND**

▲ 8.8.4 **DEMOCRACY**
Anti-Communist demonstration in Prague, Czechoslovakia, in 1990, just before the country's Communist government was replaced with a democratic one. Czechoslovakia split into two countries (Czech Republic and Slovakia) in 1993.

LOCAL GOVERNMENT: UNITARY STATE

The governments of states are organized according to one of two approaches: unitary and federal. The **unitary state** allocates most power to the national government, and local governments have relatively few powers. In principle, the unitary government system works best in nation-states characterized by few internal cultural differences and a strong sense of national unity. Because the unitary system requires effective communications with all regions of the country, smaller states are more likely to adopt it. Unitary states are especially common in Europe (Figure 8.8.5).

Some multinational states have adopted unitary systems, so that the values of one nationality can be imposed on others. In Kenya and Rwanda, for instance, the mechanisms of a unitary state have enabled one ethnic group to extend dominance over weaker groups. When Communist parties controlled the governments, most Eastern European states had unitary systems so as to promote the diffusion of Communist values.

▲ 8.8.5 **UNITARY STATE**
Monaco.

LOCAL GOVERNMENT: FEDERAL STATE

In a **federal state**, strong power is allocated to units of local government within the country. In a federal state, such as the United States, local governments possess more authority to adopt their own laws. Multinational states may adopt a federal system of government to empower different nationalities, especially if they live in separate regions of the country. Under a federal system, local government boundaries can be drawn to correspond with regions inhabited by different ethnicities.

The federal system is also more suitable for very large states because the national capital may be too remote to provide effective control over isolated regions. Most of the world's largest states are federal, including Russia (as was the former Soviet Union), Canada, the United States, Brazil, and India. However, the size of the state is not always an accurate predictor of the form of government: tiny Belgium is a federal state (to accommodate the two main cultural groups, the Flemish and the Waloons, as discussed in Chapter 5), whereas China is a unitary state (to promote Communist values).

In recent years there has been a strong global trend toward federal government. Unitary systems have been sharply curtailed in a number of countries and scrapped altogether in others (Figure 8.8.6).

▼ 8.8.6 **FEDERAL STATE**
Town hall meeting in Florida.

8.9 Electoral Geography

▶ **Gerrymandering is the drawing of legislative boundaries to favor the party in power.**

▶ **Some U.S. states gerrymandered electoral districts.**

The boundaries separating legislative districts within the United States and other countries are redrawn periodically to ensure that each district has approximately the same population. Boundaries must be redrawn because migration inevitably results in some districts gaining population, whereas others are losing. The districts of the 435 U.S. House of Representatives are redrawn every 10 years following the release of official population figures by the Census Bureau.

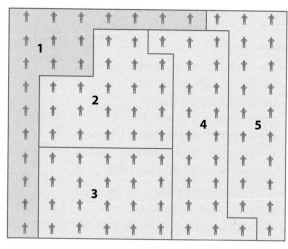

▲ 8.9.2 **"WASTED VOTE" GERRYMANDERING**
"Wasted vote" spreads opposition supporters across many districts as a minority. If the Blue Party controls the redistricting process, it could do a "wasted vote" gerrymander by creating four districts with a slender majority of Blue Party voters and one district (#1) with a strong majority of Red Party voters.

The process of redrawing legislative boundaries for the purpose of benefiting the party in power is called **gerrymandering**. The term gerrymandering was named for Elbridge Gerry (1744–1814), governor of Massachusetts (1810–12) and vice president of the United States (1813–14). As governor, Gerry signed a bill that redistricted the state to benefit his party. An opponent observed that an oddly shaped new district looked like a "salamander," whereupon another opponent responded that it was a "gerrymander." A newspaper subsequently printed an editorial cartoon of a monster named "gerrymander" with a body shaped like the district (Figure 8.9.1).

Gerrymandering works like this: suppose a community has 100 voters to be allocated among five districts of 20 voters each. The Blue Party has 52 supporters or 52 percent of the total, and the Red Party has 48 supporters or 48 percent. Gerrymandering takes three forms: "wasted vote" (Figure 8.9.2), "excess vote" (Figure 8.9.3), or "stacked vote" (Figure 8.9.4).

The job of redrawing boundaries in most European countries is entrusted to independent commissions. Commissions typically try to create compact homogeneous districts without regard for voting preferences or incumbents. A couple of U.S. states, including Iowa and Washington, also use independent or bipartisan commissions (Figure 8.9.5), but in most U.S. states the job of redrawing boundaries is entrusted to the state legislature. The political party in control of the state legislature naturally attempts to redraw boundaries to improve the chances of its supporters to win seats.

The U.S. Supreme Court ruled gerrymandering illegal in 1985 but did not require dismantling of existing oddly shaped districts, and a 2001 ruling allowed North Carolina to add another oddly shaped district that ensured the election of an African American Democrat. Through gerrymandering, only about one-tenth of Congressional seats are competitive, making a shift of more than a few seats unlikely from one election to another in the United States except in unusual circumstances.

Boundaries must be redrawn every ten years after release of census data to assure that the population is the same in each district. Political parties may offer competing plans designed to favor their candidates (Figure 8.9.6).

▼ 8.9.1 **THE ORIGINAL GERRYMANDER CARTOON**
It was drawn in 1812 by Elkanah Tinsdale.

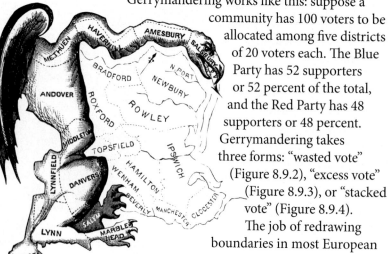

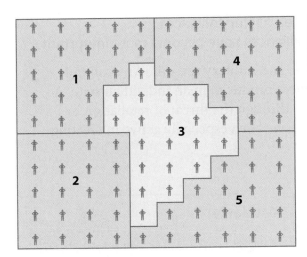

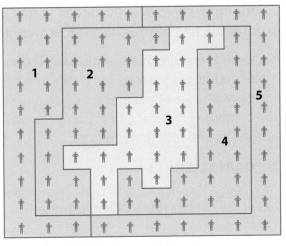

▲ 8.9.3 **"EXCESS VOTE" GERRYMANDERING**

"Excess vote" concentrates opposition supporters into a few districts. If the Red Party controls the redistricting process, it could do an "excess vote" gerrymander by creating four districts with a slender majority Red Party voters and one district (#3) with an overwhelming majority of Blue Party voters.

▲ 8.9.4 **"STACKED VOTE" GERRYMANDERING**

A "stacked vote" links distant areas of like-minded voters through oddly shaped boundaries. In this example, Red Party controls redistricting and creates five oddly shaped districts, four with a slender majority Red Party voters and one (#3) with an overwhelming majority of Blue Party voters.

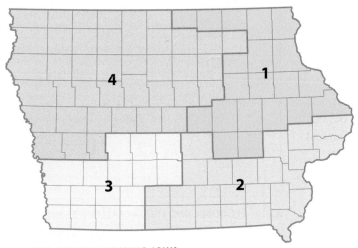

▲ 8.9.5 **NO GERRYMANDERING: IOWA**

Iowa does not have gerrymandered congressional districts. Each district is relatively compact, and boundaries coincide with county boundaries.

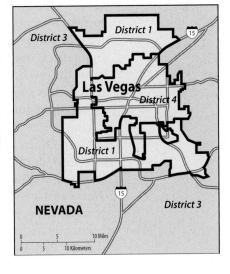

Registered voters:
- Majority Democratic
- Majority Republican

► 8.9.6 **GERRYMANDERING NEVADA: TWO PROPOSALS**

Competing proposals to draw boundaries for Nevada's four congressional districts illustrate all three forms of gerrymandering.

(top right) "Wasted vote" gerrymander. Although Nevada as a whole has slightly more registered Democrats than Republicans (43 percent to 37 percent), the Democratic plan made Democrats more numerous than Republicans in three of the four districts.

(bottom) "Excess vote" gerrymander. By clustering a large share of the state's registered Democrats in District 4, the Republican plan gave Republicans the majority of registered voters in two of the four districts.

(both) "Stacked vote" gerrymander. In the Republican plan (right), District 4 has a majority Hispanic population, and is surrounded by a "C" shaped District 1. The Democratic plan created a long, narrow District 3.

8.10 Cooperation Among States

▶ During the Cold War, European states joined military alliances.
▶ With the end of the Cold War, economic alliances have become more important.

States cooperate with each other for economic and military reasons. An economic alliance enlarges markets for goods and services produced in an individual state. A military alliance offers protection to one state through the threat of retaliation by the combined force of allies. European states have been especially active in creating economic and military alliances.

MILITARY ALLIANCES

After World War II, most European states joined one of two military alliances dominated by the superpowers:

- North Atlantic Treaty Organization (NATO): The United States, 14 Western European allies, and Canada.
- Warsaw Pact: The Soviet Union and six Eastern European allies.

In a Europe no longer dominated by military confrontation between two blocs, the Warsaw Pact and NATO became obsolete. The number of troops under NATO command was sharply reduced, and the Warsaw Pact was disbanded. Rather than disbanding, NATO expanded its membership to include most of the former Warsaw Pact countries. Membership in NATO offers eastern European countries an important sense of security against any future Russian threat, no matter how remote that appears at the moment, as well as participation in a common united Europe (Figure 8.10.1).

ECONOMIC COOPERATION

With the decline in the military-oriented alliances, European states increasingly have turned to economic cooperation. Europe's most important economic organization is the European Union. When it was established in 1958, the predecessor to the European Union included six countries. It has expanded to 12 countries during the 1980s and 27 countries during the first decade of the twenty-first century. Others hope to join.

In 1949, during the Cold War, the seven Eastern European Communist states in the Warsaw Pact formed an organization for economic cooperation, the Council for Mutual Economic Assistance (COMECON). Like the Warsaw Pact, COMECON disbanded in the

▲ 8.10.1 ECONOMIC AND MILITARY ALLIANCES IN EUROPE
Launch MapMaster Europe in Mastering GEOGRAPHY
Select *Geopolitical*, then *Economic and Military Alliances*.
Select *Political*, then *Countries*.
Select *Geopolitical*, then *Geopolitical Issues*.

Which current members of NATO and the European Union were once members of the Communist-oriented Warsaw Pact?

early 1990s after the fall of communism in Eastern Europe.

The European Union has removed most barriers to free trade. With a few exceptions, goods, services, capital, and people can move freely through Europe. A European Parliament is elected by the people in each of the member states simultaneously. Subsidies are provided to Europe's most economically depressed regions.

SUPERPOWERS

Balance of power is a condition of roughly equal strength between opposing forces. During the Cold War era (late 1940s until early 1990s), the balance of power was maintained by two superpowers—the United States and the Soviet Union (Figure 8.10.2). With the end of the Cold War, military alliances in the twenty-first century are less clearly defined (Figures 8.10.3 and 8.10.4).

▲ 8.10.3 **U. S. MILITARY VEHICLES AND AIRCRAFT LINED UP ON THE TAXIWAY AT CAMP SPEICHER, IRAQ**

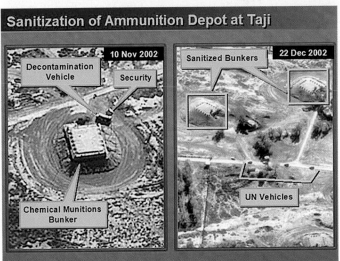

▲ 8.10.2 **COLD WAR: CUBAN MISSILE CRISIS**

A major confrontation during the Cold War between the United States and Soviet Union came in 1962 when the Soviet Union secretly began to construct missile launching sites in Cuba, less than 150 kilometers (90 miles) from U.S. territory. President Kennedy went on national television to demand that the missiles be removed and ordered a naval blockade to prevent further Soviet material from reaching Cuba.

The U.S. Department of Defense took aerial photographs to show the Soviet buildup in Cuba. (top) Three Soviet ships with missile equipment are being unloaded at Mariel naval port in Cuba. Within the outline box (enlarged below and rotated 90° clockwise) are Soviet missile transporters, fuel trailers, and oxider trailers (used to support the combustion of missile fuel).

At the United Nations, immediately after Soviet Ambassador Valerian Zorin denied that his country had placed missiles in Cuba, U.S. Ambassador Adlai Stevenson dramatically revealed aerial photographs taken by the U.S. Department of Defense clearly showing them. Faced with irrefutable evidence that the missiles existed, the Soviet Union ended the crisis by dismantling them.

▲ 8.10.4 **POST-COLD WAR: IRAQ'S ALLEGED WEAPONS**

U.S. Secretary of State Colin Powell spoke at the United Nations in 2003. The speech was supposed to present irrefutable evidence that military action against Iraq by the United States and its allies was justified. Recalling the Cuban missile crisis, Powell displayed a series of aerial photos designed to prove that Iraq possessed weapons of mass destruction. Powell first showed an image of 15 munitions bunkers at Taji, Iraq (top). He also showed close-ups of some of the bunkers (bottom).

Unlike the Cuban missile crisis in 1962, the United States could not make a convincing argument using aerial photos. As a result, the United States went to war against Iraq without the support of the United Nations. A subsequent U.S. State Department analysis found many inaccuracies in the interpretation of aerial photos presented by Powell. For example, the "decontamination vehicle" in the bottom left photo turned out to be a water truck. Two years later, Powell himself said that the 2003 speech had been a "blot" on his record.

8.11 Terrorism by Individuals and Organizations

▶ **Terrorists have attacked the United States several times.**
▶ **Al-Qaeda justifies terrorism as a holy war.**

Terrorism is the systematic use of violence by a group in order to intimidate a population or coerce a government into granting its demands. Terrorists attempt to achieve their objectives through organized acts that spread fear and anxiety among the population, such as bombing, kidnapping, hijacking, taking of hostages, and assassination. They consider violence necessary as a means of bringing widespread publicity to goals and grievances that are not being addressed through peaceful means. Belief in their cause is so strong that terrorists do not hesitate to strike despite knowing they will probably die in the act.

TERRORISM AGAINST AMERICANS

The most dramatic terrorist attack against the United States came on September 11, 2001. The tallest buildings in the United States, the 110-story twin towers of the World Trade Center in New York City were destroyed, and the Pentagon in Washington, D.C., was damaged (Figure 8.11.1). The attacks resulted in nearly 3,000 fatalities.

Prior to the 9/11 attacks, the United States had suffered several terrorist attacks during the late

▲ 8.11.1 **SEPTEMBER 11, 2001 ATTACKS**

twentieth century (Figure 8.11.2). Some were by American citizens operating alone or with a handful of others.

• Theodore J. Kaczynski, known as the Unabomber, was convicted of killing 3 people and injuring 23 others by sending bombs through the mail during a 17-year period. His targets were mainly academics in technological disciplines and executives in businesses whose actions he considered to be adversely affecting the environment.

• Timothy J. McVeigh was convicted and executed for the Oklahoma City bombing, and for assisting him Terry I. Nichols was convicted of conspiracy and involuntary manslaughter. McVeigh claimed he had been provoked by U.S. government actions including the FBI's 51-day siege of the Branch Davidian religious compound near Waco, Texas, culminating with an attack on April 19, 1993, that resulted in 80 deaths.

AL-QAEDA

Responsible or implicated in most of the anti-U.S. terrorism in Figure 8.11.2, including the September 11, 2001, attack, was the al-Qaeda network (Figure 8.11.3). Al-Qaeda (an Arabic word meaning "the foundation" or "the base") has

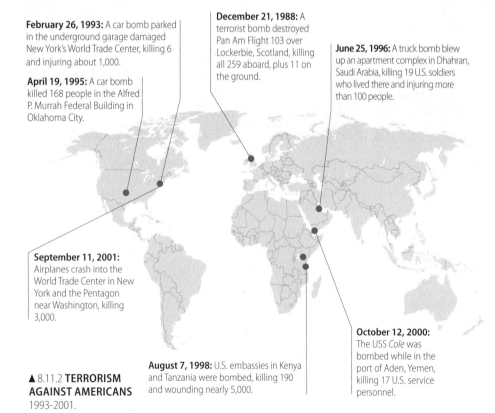

February 26, 1993: A car bomb parked in the underground garage damaged New York's World Trade Center, killing 6 and injuring about 1,000.

April 19, 1995: A car bomb killed 168 people in the Alfred P. Murrah Federal Building in Oklahoma City.

December 21, 1988: A terrorist bomb destroyed Pan Am Flight 103 over Lockerbie, Scotland, killing all 259 aboard, plus 11 on the ground.

June 25, 1996: A truck bomb blew up an apartment complex in Dhahran, Saudi Arabia, killing 19 U.S. soldiers who lived there and injuring more than 100 people.

September 11, 2001: Airplanes crash into the World Trade Center in New York and the Pentagon near Washington, killing 3,000.

October 12, 2000: The USS *Cole* was bombed while in the port of Aden, Yemen, killing 17 U.S. service personnel.

August 7, 1998: U.S. embassies in Kenya and Tanzania were bombed, killing 190 and wounding nearly 5,000.

▲ 8.11.2 **TERRORISM AGAINST AMERICANS** 1993-2001.

Interactive Mapping

TERRITORIAL CLAIMS IN THE SOUTH PACIFIC

Microstates in the Pacific Ocean have tiny areas but they control vast areas of the sea.

Launch MapMaster Australia and Oceania in Mastering**GEOGRAPHY**

Select: *Geopolitical* then *Marine Tropical Claims*.
Select: *Political* then *Countries*.

1. **How many independent microstates are named in the region (countries in addition to Australia, New Zealand, and Papua New Guinea)?**

2. **How many states have colonies in the South Pacific?**

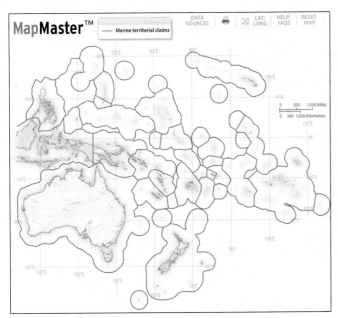

Explore

THE KREMLIN, MOSCOW, RUSSIA

Use Google Earth to explore the Kremlin, the fortified center of Moscow that has symbolized the power of the Soviet Union and now Russia.

Fly to: *The Kremlin, Moscow, Russia*

Click 3D buildings.

Drag to enter Street View in the large square in the center of the image.

Exit Ground Level View and zoom out until the walls of the complex are visible.

Click on the individual buildings.

What are the current use and history of the major buildings in the Kremlin?

Key Terms

Anocracy
A country that is not fully democratic or fully autocratic, but rather displays a mix of the two types.

Autocracy
A country that is run according to the interests of the ruler rather than the people.

Balance of power
Condition of roughly equal strength between opposing countries or alliances of countries.

Boundary
Invisible line that marks the extent of a state's territory.

City-state
A sovereign state comprising a city and its immediate hinterland.

Colonialism
Attempt by one country to establish settlements and to impose its political, economic, and cultural principles in another territory.

Colony
A territory that is legally tied to a sovereign state rather than completely independent.

Compact state
A state in which the distance from the center to any boundary does not vary significantly.

Elongated state
A state with a long, narrow shape.

Federal state
An internal organization of a state that allocates most powers to units of local government.

Fragmented state
A state that includes several discontinuous pieces of territory.

Frontier
A zone separating two states in which neither state exercises political control.

Gerrymandering
Process of redrawing legislative boundaries for the purpose of benefiting the party in power.

Landlocked state
A state that does not have a direct outlet to the sea.

Multinational state
State that contains two or more ethnic groups with traditions of self-determination that agree to coexist peacefully by recognizing each other as distinct nationalties.

Nation-state
A state whose territory corresponds to that occupied by a particular ethnicity that has been transformed into a nationality.

Perforated state
A state that completely surrounds another one.

Prorupted state
An otherwise compact state with a large projecting extension.

Self-determination
Concept that ethnicities have the right to govern themselves.

Sovereignty
Ability of a state to govern its territory free from control of its internal affairs by other states.

State
An area organized into a political unit and ruled by an established government with control over its internal and foreign affairs.

Terrorism
The systematic use of violence by a group in order to intimidate a population or coerce a government into granting it demands.

Unitary state
An internal organization of a state that places most power in the hands of central government officials.

On the Internet

The U.S. Central Intelligence Agency has a World Factbook. Select a country from the drop-down list to find background information, as well as facts and figures about the country's demography, economy, physical geography, government, and military. Maps are also available at **https://www.cia.gov/library/publications/the-world-factbook/** or scan the QR at the beginning of the chapter.

▶ **LOOKING AHEAD**

The second half of the book concentrates on economic elements of human geography, beginning with the division of the world into more and less developed regions.

9 Development

The world is divided into developed countries and developing countries. The one-fifth of the world's people living in developed countries consume five-sixths of the world's goods, whereas the 14 percent of the world's people who live in Africa consume about 1 percent.

The United Nations recently contrasted spending between developed and developing countries in picturesque terms: Americans spend more per year on cosmetics ($8 billion) than the cost of providing schools for the 2 billion people in the world in need of them ($6 billion).

Europeans spend more on ice cream ($11 billion) than the cost of providing a working toilet to the 2 billion people currently without one at home ($9 billion).

To reduce disparities between rich and poor countries, developing countries must develop more rapidly. This means increasing wealth and using that wealth to make more rapid improvements in people's health and well-being.

How does development vary among regions?

9.1 **Human Development Index**

9.2 **Standard of Living**

9.3 **Access to Knowledge**

9.4 **Health Indicators**

9.5 **Gender-related Development**

BUILDING A NEW
ROAD, MOZAMBIQUE

How can countries promote development?

What are future challenges for development?

SCAN TO ACCESS THE UN's HUMAN DEVELOPMENT REPORT

9.1 Human Development Index

▶ **Countries are classified as developed or developing.**

▶ **The Human Development Index (HDI) measures a country's level of development.**

Earth's nearly 200 countries can be classified according to their level of **development**, which is the process of improving the material conditions of people through diffusion of knowledge and technology. The development process is continuous, involving never-ending actions to constantly improve the health and prosperity of the people. Every place lies at some point along a continuum of development.

The United Nations classifies countries as developed or developing:

- A **developed country**, also known as a **more developed country (MDC)** or a **relatively developed country,** has progressed further along the development continuum. The UN considers these countries to have very high development.

- A **developing country**, also frequently called a **less developed country (LDC),** has made some progress towards development though less than developed countries. Recognizing that progress has varied widely among developing countries, the UN divides them into high, medium, and low development.

To measure the level of development of every country, the UN created the **Human Development Index (HDI)**. The UN has computed HDIs for countries every year since 1990, although it has occasionally modified the method of computation. The HDI considers development to be a function of three factors:
- A decent standard of living.
- Access to knowledge.
- A long and healthy life.

Each country gets a score for each of these three factors, which are then combined into an overall HDI (Figure 9.1.1). The highest HDI possible is 1.0, or 100 percent. These factors are discussed in more detail in sections 9.2, 9.3, and 9.4.

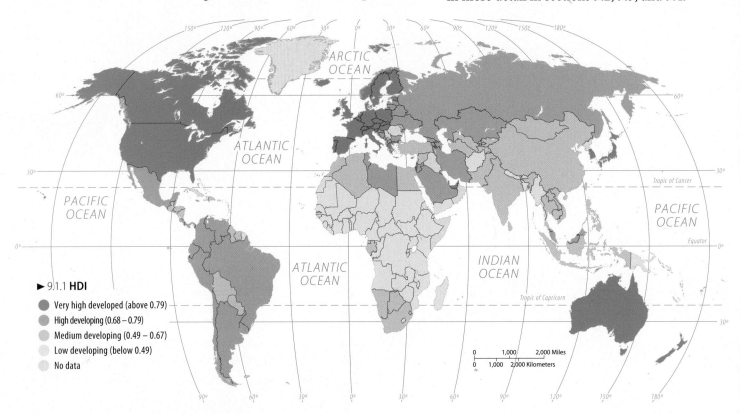

▶ **9.1.1 HDI**

- Very high developed (above 0.79)
- High developing (0.68 – 0.79)
- Medium developing (0.49 – 0.67)
- Low developing (below 0.49)
- No data

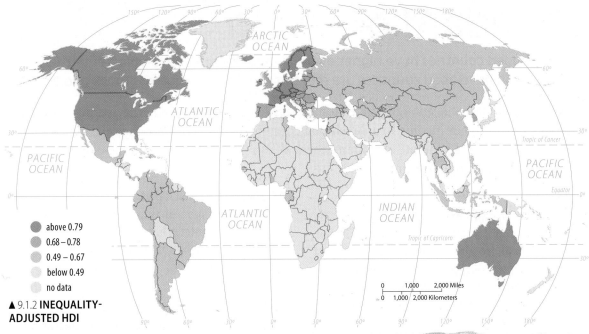

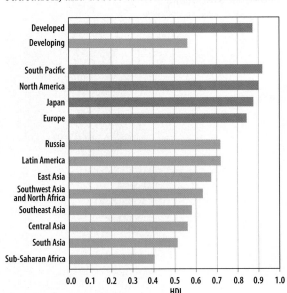

above 0.79
0.68 – 0.78
0.49 – 0.67
below 0.49
no data

▲ 9.1.2 **INEQUALITY-ADJUSTED HDI**

INEQUALITY-ADJUSTED HDI

The United Nations believes that every person should have access to decent standards of living, knowledge, and health. The **Inequality-adjusted HDI (IHDI)** modifies the HDI to account for inequality (Figure 9.1.2).

Under perfect equality the HDI and the IHDI are the same. If the IHDI is lower than the HDI, the country has some inequality; the greater the difference in the two measures, the greater the inequality. A country where only a few people have high incomes, college degrees, and good health care would have a lower IHDI than a country where differences in income, level of education, and access to health care are minimal.

Europe
North America
Latin America
Southwest Asia and North Africa
Sub-Saharan Africa
Central Asia
East Asia
South Asia
Southeast Asia

Japan
Russia
South Pacific

▲ 9.1.3 **NINE WORLD REGIONS**

FOCUS ON WORLD REGIONS

Geographers divide the world into nine regions according to physical, cultural, and economic features (Figure 9.1.3). Two of the nine regions—North America and Europe—are considered developed (Figure 9.1.4). The other seven regions—Latin America, East Asia, Southwest Asia & North Africa, Southeast Asia, Central Asia, South Asia, and sub-Saharan Africa—are considered developing. In addition to these nine regions, three other distinctive areas can be identified—Japan, Russia, and South Pacific. Japan and South Pacific are grouped with the developed regions. Because of limited progress in development both under and since communism, Russia is now classified as a developing country by the United Nations. In each of the remaining nine sections of this chapter, one of the nine regions is highlighted in relation to the topic of the section.

▲ 9.1.4 **HDI BY REGION**

(Bar chart: Developed, Developing, South Pacific, North America, Japan, Europe, Russia, Latin America, East Asia, Southwest Asia and North Africa, Southeast Asia, Central Asia, South Asia, Sub-Saharan Africa; x-axis HDI from 0.0 to 1.0)

9.2 Standard of Living

▶ **Developed countries have higher average incomes than developing countries.**
▶ **People in developed countries are more productive and possess more goods.**

Key to development is enough wealth for a decent standard of living. The average individual earns a much higher income in a developed country than in a developing one. Geographers observe that people generate and spend their wealth in different ways in developed countries than in developing countries.

INCOME

The United Nations measures the average income in countries through a complex index called annual gross national income per capita at purchasing power parity. The figure is approximately $40,000 in developed countries compared to approximately $5,000 in developing countries (Figure 9.2.1).

Gross national income (GNI) is the value of the output of goods and services produced in a country in a year, including money that leaves and enters the country. Dividing GNI by total population measures the contribution made by the average individual towards generating a country's wealth in a year. Older studies refer to **gross domestic product**, which is also the value of the output of goods and services produced in a country in a year, but it does not account for money that leaves and enters the country.

Purchasing power parity (PPP) is an adjustment made to the GNI to account for differences among countries in the cost of goods. For example, if a resident of country A has the same income as a resident in country B but must pay more for a Big Mac or a Starbucks latte, the resident of country B is better off.

ECONOMIC STRUCTURE

Average per capita income is higher in developed countries because people typically earn their living by different means than in developing countries. Jobs fall into three categories:

• **Primary sector** (including agriculture).
• **Secondary sector** (including manufacturing).
• **Tertiary sector** (including services).

Developing countries have a higher share of primary and secondary sector workers and a smaller share of tertiary sector workers than developed countries (Figure 9.2.2). The relatively low percentage of primary-sector workers in developed countries indicates that a handful of farmers produce enough food for the rest of society. Freed from the task of growing their own food, most people in a developed country can contribute to an increase in the national wealth by working in the secondary and tertiary sectors (Figure 9.2.3).

▼ 9.2.2 **FOCUS ON NORTH AMERICA: ECONOMIC STRUCTURE**
Tertiary-sector workers in Florida.

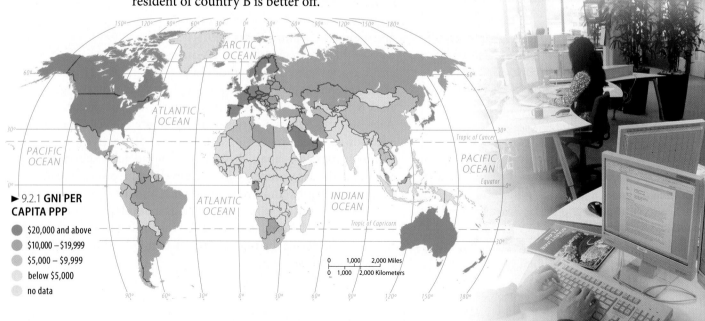

▶ 9.2.1 **GNI PER CAPITA PPP**

● $20,000 and above
● $10,000 – $19,999
● $5,000 – $9,999
● below $5,000
● no data

ARCTIC OCEAN

ATLANTIC OCEAN

PACIFIC OCEAN

ATLANTIC OCEAN

PACIFIC OCEAN

INDIAN OCEAN

Tropic of Cancer

Equator

Tropic of Capricorn

0   1,000   2,000 Miles
0   1,000   2,000 Kilometers

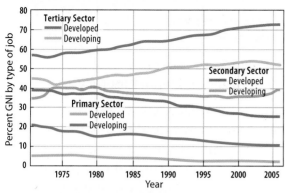

▲ 9.2.3 **PERCENT GNI CONTRIBUTED BY TYPE OF JOB**

PRODUCTIVITY

Workers in developed countries are more productive than those in developing ones. **Productivity** is the value of a particular product compared to the amount of labor needed to make it. Productivity can be measured by the value added per worker. The **value added** in manufacturing is the gross value of the product minus the costs of raw materials and energy. Workers in developed countries produce more with less effort because they have access to more machines, tools, and equipment to perform much of the work (Figure 9.2.4).

▲ 9.2.4 **FOCUS ON NORTH AMERICA: PRODUCTIVITY** Manufacturing computers in California.

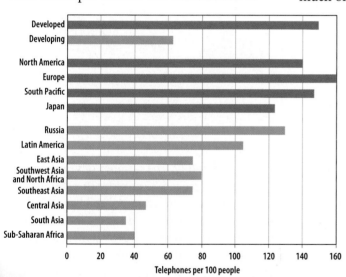

▲ 9.2.5 **TELEPHONES PER 100 PEOPLE**

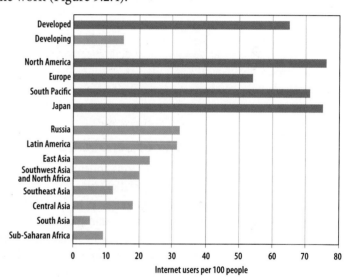

▲ 9.2.6 **INTERNET USERS PER 100 PEOPLE**

CONSUMER GOODS

Part of the wealth generated in developed countries is used to purchase goods and services. Especially vital to the economy's functioning and growth are goods and services related to communications, such as telephones and computers. Computers and telephones are not essential to people who live in the same village as their friends and relatives and work all day growing food in nearby fields.

Telephones enhance interaction with providers of raw materials and customers for goods and services (Figure 9.2.5). Computers facilitate the sharing of information with other buyers and suppliers (Figure 9.2.6). Developed countries average 150 telephones and 65 Internet users per 100 persons, compared to 60 telephones and 15 Internet users per 100 in developing countries.

FOCUS ON NORTH AMERICA

North America is the region with the world's highest per capita income. North America was once the world's major manufacturer of steel, motor vehicles, and other goods, but since the late twentieth century other regions have taken the lead. Now the region has the world's highest percentage of tertiary-sector employment, especially health care, leisure, and financial services. North Americans remain the leading consumers and the world's largest market for many products. The wealth generated in the United States and Canada enables the residents of those countries to purchase more consumer goods than in other regions.

9.3 Access to Knowledge

▶ **People in developed countries complete more years of school.**
▶ **Developed countries have lower pupil/teacher ratios and higher literacy.**

Development is about more than possession of wealth. The United Nations believes that access to knowledge is essential for people to have the possibility of leading lives of value. In general, the higher the level of development, the greater are both the quantity and the quality of a country's education. For many in developing countries, education is the ticket to better jobs and higher social status.

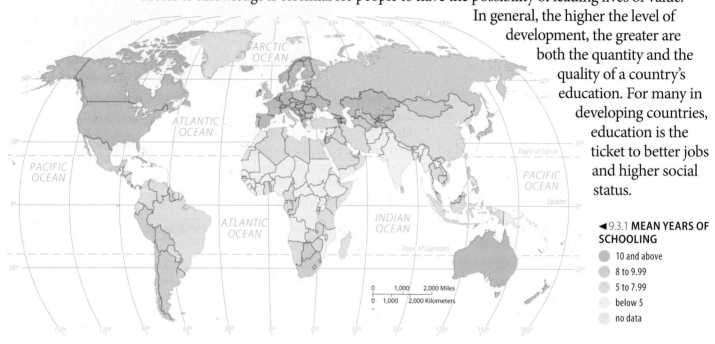

◀ 9.3.1 **MEAN YEARS OF SCHOOLING**
- 10 and above
- 8 to 9.99
- 5 to 7.99
- below 5
- no data

QUANTITY OF SCHOOLING

The United Nations considers years of schooling to be the most critical measure of the ability of an individual to gain access to knowledge needed for development. The assumption is that no matter how poor the school, the longer the pupils attend, the more likely they are to learn something.

To form the access to knowledge component of HDI, the United Nations combines two measures of quantity of schooling:

▼ 9.3.2 **EXPECTED YEARS OF SCHOOLING**

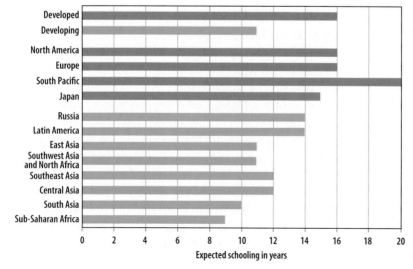

Expected schooling in years

- *Years of schooling.* The number of years that the average person aged 25 or older in a country has spent in school. The average pupil has attended school for approximately 11 years in developed countries, compared to approximately 6 years in developing countries (Figure 9.3.1).

- *Expected years of schooling.* The number of years that an average 5 year old child is expected to spend with his or her education in the future. The United Nations expects that today's 5-year-old will attend an average of 16 years of school in developed countries and 11 years in developing ones (Figure 9.3.2). Sub-Saharan Africa and South Asia are expected to lag in schooling compared to other regions.

Thus, the United Nations expects children around the world to receive an average of five years more education in the future, but the gap in education between developed and developing regions will remain high. Otherwise stated, the United Nations expects that roughly half of today's 5-year-olds will graduate from college in developed countries, whereas less than half will graduate from high school in developing ones.

QUALITY OF SCHOOLING

Two measures of quality of education include:

• *Pupil/teacher ratio.* The fewer pupils a teacher has, the more likely that each student will receive instruction. The pupil/teacher ratio is twice as high in developing countries—approximately 30 pupils per teacher—compared to only 15 in developed countries (Figure 9.3.3). Pupil/teacher ratio exceeds 40 in sub-Saharan Africa and South Asia.

• *Literacy rate.* A higher percentage of people in developed countries are able to attend school and as a result learn to read and write. The **literacy rate** is the percentage of a country's people who can read and write. It exceeds 99 percent in developed countries (Figure 9.3.4). Among developing regions, the literacy rate exceeds 90 percent in East Asia and Latin America, but is less than 70 percent in sub-Saharan Africa and South Asia.

Most books, newspapers, and magazines are published in developed countries, in part because more of their citizens read and write. Developed countries dominate scientific and nonfiction publishing worldwide—this textbook is an example. Students in developing countries must learn technical information from books that usually are not in their native language but are printed in English, German, Russian, or French.

Improved education is a major goal of many developing countries, but funds are scarce. Education may receive a higher percentage of the GNI in developing countries, but their GNI is far lower to begin with, so they spend far less per pupil than do developed countries.

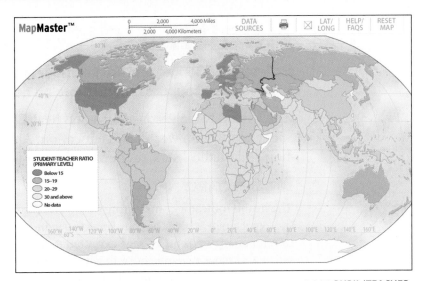

▲ 9.3.3 **PUPIL/TEACHER RATIO**

Open MapMaster World in Mastering**GEOGRAPHY**

Select: *Cultural* then *Students per teacher in primary school.*
Select: *Population* then *Percentage of population under age 15.*

Are class sizes larger or smaller in countries that have a high percentage of population under age 15?

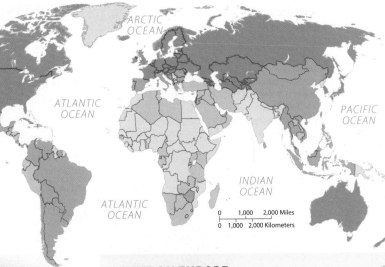

► 9.3.4 **LITERACY RATE**

Percent literate
- 99–100
- 90–98
- 70–89
- below 70
- no data

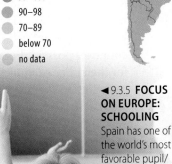

◄ 9.3.5 **FOCUS ON EUROPE: SCHOOLING**
Spain has one of the world's most favorable pupil/teacher ratios.

FOCUS ON EUROPE

Within Europe, the HDI is the world's highest in a core area that extends from southern Scandinavia to western Germany. These countries have especially high levels of schooling, favorable pupil/teacher ratios, and universal literacy (Figure 9.3.5). Europe's overall development indicators are somewhat lower because of inclusion of Eastern European countries that developed under communist rule for much of the twentieth century. Europe must import food, energy, and minerals, but can maintain its high level of development by providing high value goods and services, such as insurance, banking, and luxury motor vehicles.

9.4 Health Indicators

▶ **People live longer and are healthier in developed countries.**
▶ **Developed countries spend more on health care.**

The United Nations considers good health to be a third important measure of development, along with wealth and education. A goal of development is to provide the nutrition and medical services needed for people to lead long and healthy lives.

LIFE EXPECTANCY

The health indicator contributing to the HDI is life expectancy at birth. A baby born today in a developed region is on average expected to live ten years longer than one born in a developing region (Figure 9.4.1, and refer to Figure 2.4.3 for world map). Variation among developing regions is especially wide; life expectancy in East Asia and Latin America is comparable to the level in developed countries, but it is much lower in sub-Saharan Africa.

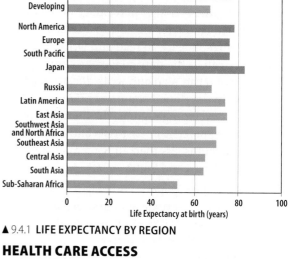

▲ 9.4.1 **LIFE EXPECTANCY BY REGION**

HEALTH CARE ACCESS

People live longer and are healthier in developed countries than in developing ones because of better access to health care. The greater wealth that is generated in developed countries is used in part to obtain health care. A healthier population in turn can be more economically productive. For example, 17 percent of children in developing countries are not immunized against measles, compared to 7 percent in developed ones. More than one-fourth of children lack measles immunization in South Asia and sub-Saharan Africa (Figure 9.4.2).

When people get sick, developed countries possess the resources to care for them. For example, developed countries on average have 50 hospital beds per 10,000 population compared to only 20 in developing countries (Figures 9.4.3 and 9.4.4).

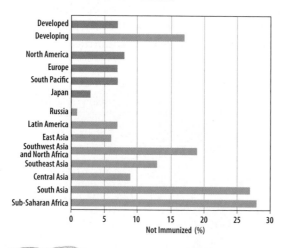

▶ 9.4.2 **CHILDREN LACKING MEASLES IMMUNIZATION**

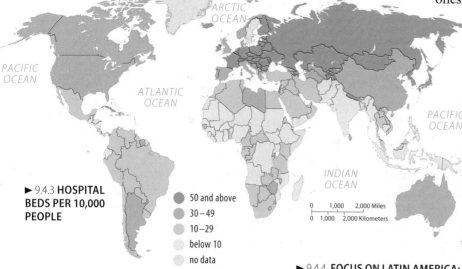

▶ 9.4.3 **HOSPITAL BEDS PER 10,000 PEOPLE**

- 50 and above
- 30 – 49
- 10 – 29
- below 10
- no data

▶ 9.4.4 **FOCUS ON LATIN AMERICA: HEALTH CARE**
Clinic in Haiti run by American missionaries.

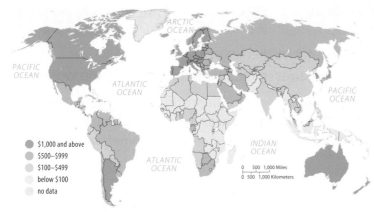

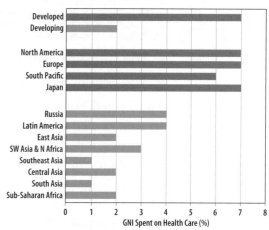

▲ 9.4.5 **HEALTH CARE EXPENDITURE PER CAPITA**

▲ 9.4.6 **HEALTH CARE EXPENDITURE AS PERCENTAGE OF GNI**

HEALTH CARE EXPENDITURES

The gap between developed and developing countries is especially high in expenditures on health care. Developed countries spend more than $4,000 per person annually on health care, compared to approximately $200 per person in developing countries (Figure 9.4.5). Hospitals, medicines, doctors—spending is much higher in developed countries.

Total expenditures on health care exceed 7 percent of GNI in developed countries, compared to 2 percent in developing ones. So not only do developed countries have much higher GNI per capita than developing countries, they spend a higher percentage of that GNI on health care (Figure 9.4.6).

In most developed countries, health care is a public service that is available at little or no cost. The government programs pay more than 70 percent of health care costs in most European countries, and private individuals pay less than 30 percent. In comparison, private individuals must pay more than half of the cost of health care in developing countries. An exception is the United States, where private individuals are required to pay 55 percent of health care, more closely resembling the pattern in developing countries.

Developed countries also use part of their wealth to protect people who, for various reasons, are unable to work. In these countries some public assistance is offered to those who are sick, elderly, poor, disabled, orphaned, veterans of wars, widows, unemployed, or single parents. European countries such as Denmark, Norway, and Sweden typically provide the highest level of public-assistance payments.

Developed countries are hard-pressed to maintain their current levels of public assistance. In the past, rapid economic growth permitted these states to finance generous programs with little hardship. But in recent years economic growth has slowed, whereas the percentage of people needing public assistance has increased. Governments have faced a choice between reducing benefits or increasing taxes to pay for them.

▼ 9.4.7 **FOCUS ON LATIN AMERICA: HEALTH CARE** Clinic in Colombia for displaced people.

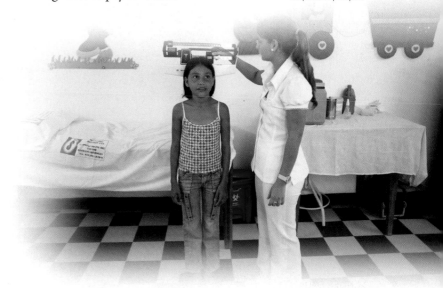

FOCUS ON LATIN AMERICA

The level of development varies sharply within Latin America. Neighborhoods within some large cities along the South Atlantic Coast enjoy a level of development comparable to that of developed countries. The coastal area as a whole has a relatively high GNI per capita. Outside the coastal area, development is lower. Among developing regions, Latin America—along with East Asia—has relatively high life expectancy, high immunization rates, more hospital beds per capita, and more money spent on health care. The levels lag, though, compared with developed regions.

9.5 Gender-Related Development

▶ **The status of women is lower than that of men in every country.**

▶ **The Gender Inequality Index (GII) measures inequality between men and women.**

The United Nations has not found a single country in the world where women are treated as well as men. At best women have achieved near equality with men in some countries, whereas in other countries the level of development of women lags far behind the level for men.

To measure the extent of each country's gender inequality, the United Nations has created the **Gender Inequality Index (GII)**. The higher the score the greater is the inequality between men and women (Figure 9.5.1). As with the other indices, the GII combines multiple measures, in this case reproductive health, empowerment, and labor.

EMPOWERMENT

The empowerment dimension is measured by two indicators:

• The percentage of seats held by women in the national legislature (Figures 9.5.2 and 9.5.3).

• The percentage of women who have completed high school.

Both measures are lower in developing regions than in developed ones.

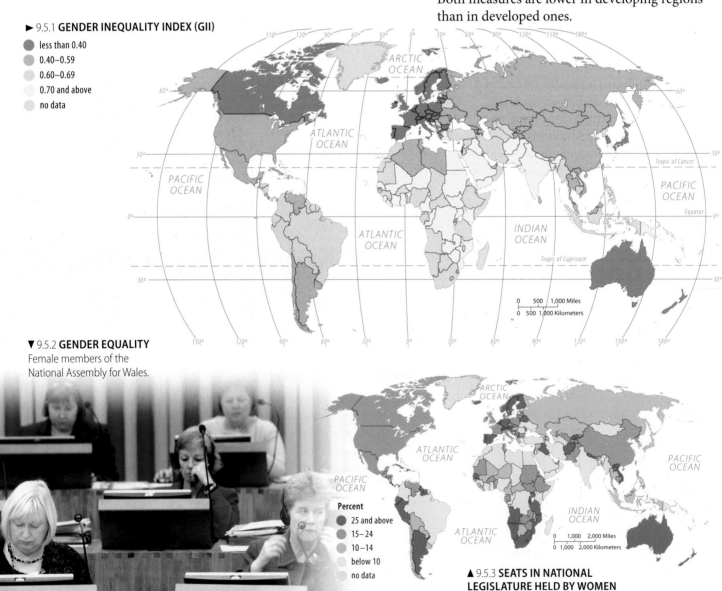

▶ 9.5.1 **GENDER INEQUALITY INDEX (GII)**
- less than 0.40
- 0.40–0.59
- 0.60–0.69
- 0.70 and above
- no data

▼ 9.5.2 **GENDER EQUALITY**
Female members of the National Assembly for Wales.

Percent
- 25 and above
- 15–24
- 10–14
- below 10
- no data

▲ 9.5.3 **SEATS IN NATIONAL LEGISLATURE HELD BY WOMEN**

LABOR

The labor force participation rate is the percent of women holding full-time jobs outside the home. Women in developing countries are less likely than women in developed countries to hold full-time jobs outside the home (Figure 9.5.4).

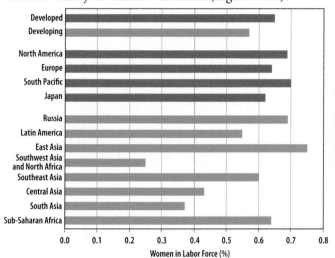

Women in Labor Force (%)

▲ 9.5.4 **WOMEN IN LABOR FORCE**
▼ 9.5.5 **ADOLESCENT FERTILITY RATE (right)**
TEENAGE MOTHER IN OHIO (below)

▲ 9.5.6 **FOCUS ON EAST ASIA: WOMEN IN THE LABOR FORCE**
Female workers in optical fiber factory, Ghuanzhou, China.

Births per 1,000 aged 15–19
- below 15
- 15–39
- 40–79
- 80 and above
- no data

REPRODUCTIVE HEALTH

The health dimension is also measured by two indicators:

- **Maternal mortality ratio** is the number of women who die giving birth per 100,000 births.
- **Adolescent fertility rate** is the number of births per 1,000 women age 15–19.

Women in developing regions are more likely than women in developed regions to die in childbirth and to give birth as teenagers (Figure 9.5.5).

In general, the GII is higher in developing regions than in developed ones. Sub-Saharan Africa, South Asia, Central Asia, and Southwest Asia are the developing regions with the highest levels of gender inequality. Reproductive health is the largest contributor to gender inequality in these regions. South and Southwest Asia also have relatively poor female empowerment scores. The United States ranks especially poorly in the percentage of teenagers who give birth and in the percentage of women serving in Congress.

FOCUS ON EAST ASIA

The Gender Inequality Index in East Asia is comparable to that of developed regions. Compared to other developing countries, China has high female education levels and participation in the labor force and low maternal mortality and teenage fertility rates (Figure 9.5.6). Now the world's second largest economy, behind only the United States, China accounts for one-third of total world economic growth, and GNI per capita has risen faster there than in any other country. Under communism, the government took strong control of most components of development.

9.6 Two Paths to Development

▶ **The self-sufficiency development path erects barriers to trade.**
▶ **The international trade path allocates scarce resources to a few activities.**

To promote development, developing countries typical follow one of two development models. One emphasizes self-sufficiency, the other international trade.

DEVELOPMENT THROUGH SELF-SUFFICIENCY

Self-sufficiency, or balanced growth, was the more popular of the development alternatives for most of the twentieth century. According to the self-sufficiency approach:

- Investment is spread as equally as possible across all sectors of a country's economy and in all regions.
- The pace of development may be modest, but the system is fair because residents and enterprises throughout the country share the benefits of development.
- Reducing poverty takes precedence over encouraging a few people to become wealthy consumers.
- Fledgling businesses are isolated from competition with large international corporations.
- The import of goods from other places is limited by barriers such as tariffs, quotas, and licenses.

SELF-SUFFICIENCY EXAMPLE: INDIA

India once followed the self-sufficiency model (Figure 9.6.1). India's barriers to trade included:

- To import goods into India, most foreign companies had to secure a license that had to be approved by several dozen government agencies.
- An importer with a license was severely restricted in the quantity it could sell in India.
- Heavy taxes on imported goods doubled or tripled the price to consumers.
- Indian money could not be converted to other currencies.
- Businesses required government permission to sell a new product, modernize a factory, expand production, set prices, hire or fire workers, and change the job classification of existing workers.

▼ 9.6.1 **SELF-SUFFICIENCY EXAMPLE: INDIA**
Basmati rice on sale at a market in Haryana.

DEVELOPMENT THROUGH INTERNATIONAL TRADE

According to the international trade approach, a country can develop economically by concentrating scarce resources on expansion of its distinctive local industries. The sale of these products in the world market brings funds into the country that can be used to finance other development. W. W. Rostow proposed a five-stage model of development in 1960.

- **The traditional society.** A very high percentage of people engaged in agriculture and a high percentage of national wealth allocated to what Rostow called "nonproductive" activities, such as the military and religion.
- **The preconditions for takeoff.** An elite group of well-educated leaders initiates investment in technology and infrastructure, such as water supplies and transportation systems, designed to increase productivity.
- **The takeoff.** Rapid growth is generated in a limited number of economic activities, such as textiles or food products.
- **The drive to maturity.** Modern technology, previously confined to a few takeoff industries, diffuses to a wide variety of industries.
- **The age of mass consumption.** The economy shifts from production of heavy industry, such as steel and energy, to consumer goods, such as motor vehicles and refrigerators.

INTERNATIONAL TRADE EXAMPLES

Among the first countries to adopt the international trade alternative during the twentieth century:

- **The "Four Dragons."** South Korea, Singapore, Taiwan, and the then-British colony of Hong Kong (also known as the "four little tigers" and "the gang of four") developed by producing a handful of manufactured goods, especially clothing and electronics, that depended on low labor costs.
- **Petroleum-rich Arabian Peninsula countries.** Once among the world's least developed countries, they were transformed overnight into some of the wealthiest thanks to escalating petroleum prices during the 1970s (Figure 9.6.2).

SELF-SUFFICIENCY SHORTCOMINGS

The experience of India and other developing countries revealed two major problems with self-sufficiency:

- **Self-sufficiency protected inefficient industries.** Businesses could sell all they made, at high government-controlled prices, to customers culled from long waiting lists. So they had little incentive to improve quality, lower production costs, reduce prices, or increase production. Nor did they keep abreast of rapid technological changes elsewhere.
- **A large bureaucracy was needed to administer the controls.** A complex administrative system encouraged abuse and corruption. Aspiring entrepreneurs found that struggling to produce goods or offer services was less rewarding financially than advising others how to get around the complex regulations.

INTERNATIONAL TRADE SHORTCOMINGS

Three factors have hindered countries outside the four Asian dragons and the Arabian Peninsula from developing through the international trade:

- **Local hardships.** Building up a handful of takeoff industries has forced some developing countries to cut back on production of food, clothing, and other necessities for their own people.
- **Slow market growth.** Developing countries trying to take advantage of their low-cost labor find that markets in developed countries are growing more slowly than when the "four dragons" used this strategy a generation ago.
- **Low commodity prices.** Some developing countries have raw materials sought by manufacturers and producers in developed countries. The sale of these raw materials could generate funds for developing countries to promote development. International trade worked in the Arabian Peninsula because the price of petroleum has escalated so rapidly, but other developing countries have not been so fortunate because of low prices for their commodities.

▲ 9.6.2 **INTERNATIONAL TRADE EXAMPLE: UNITED ARAB EMIRATES** Development in Dubai, United Arab Emirates.

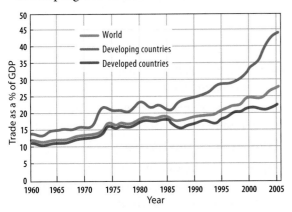

▲ 9.6.3 **WORLD TRADE AS PERCENT OF INCOME**

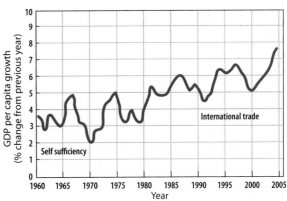

▲ 9.6.4 **GDP PER CAPITA CHANGE IN INDIA**

INTERNATIONAL TRADE TRIUMPHS

Countries have converted from self-sufficiency to international trade (Figure 9.6.3). For example, India has:

- Reduced taxes and restrictions on imports and exports
- Eliminated many monopolies
- Encouraged improvement of the quality of products

India's per capita income has increased more rapidly since conversion to international trade (Figure 9.6.4).

FOCUS ON SOUTHWEST ASIA AND NORTH AFRICA

Countries in Southwest Asia and North Africa that are oil-rich have used petroleum revenues to finance large-scale projects, such as housing, highways, airports, universities, and telecommunications networks. Imported consumer goods are readily available. However, some business practices typical of international trade are difficult to reconcile with Islamic religious principles. Women are excluded from holding many jobs and visiting some public places. All business halts several times a day when Muslims are called to prayers.

9.7 World Trade

► **The World Trade Organization has facilitated adoption of international trade.**
► **Transnational corporations are a major source of development funds.**

To promote the international trade development model, most countries have joined the World Trade Organization (WTO). Private corporations are especially eager to promote international trade.

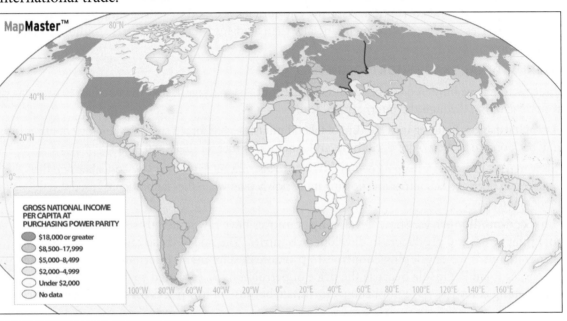

► 9.7.1 **WORLD TRADE ORGANIZATION**
Open MapMaster World in

Mastering GEOGRAPHY™

Select: *Economic* then *Gross National Income Per Capita at Purchasing Power Parity*.
Select: *Geopolitical* then *World Trade Organization Members*.

In what three of the nine major world regions (excluding Russia) are most of the countries not members of the WTO?

GROSS NATIONAL INCOME PER CAPITA AT PURCHASING POWER PARITY
- $18,000 or greater
- $8,500–17,999
- $5,000–8,499
- $2,000–4,999
- Under $2,000
- No data

WORLD TRADE ORGANIZATION

To promote the international trade development model, countries representing 97 percent of world trade established the WTO in 1995. Russia is the largest economy that has not joined the WTO (Figure 9.7.1). The WTO works to reduce barriers to trade in three principal ways:
1. Reduce or eliminate restrictions:
- On trade of manufactured goods, such as government subsidies of exports, quotas, and tariffs.
- On international movement of money by banks, corporations, and wealthy individuals.

2. Enforce agreements:
- By ruling on whether a country has violated WTO agreements.
- By ordering remedies when one country has been found to have violated the agreements.

3. Protect intellectual property:
- By hearing charges from an individual or corporation concerning copyright and patent violations in other countries.
- By ordering illegal copyright or patent activities to stop.

The WTO has been sharply attacked by critics (Figure 9.7.2). Protesters routinely gather in the streets outside high-level meetings of the WTO:
- Progressive critics charge that the WTO is antidemocratic, because decisions made behind closed doors promote the interest of large corporations rather than the poor.
- Conservative critics charge that the WTO compromises the power and sovereignty of individual countries because it can order changes in taxes and laws that it considers unfair trading practices.

▲▼ 9.7.2 **THE WORLD TRADE ORGANIZATION GENERATES STRONG SUPPORT AND OPPOSITION**

FOREIGN DIRECT INVESTMENT

International trade requires corporations based in a particular country to invest in other countries (Figure 9.7.3). Investment made by a foreign company in the economy of another country is known as **foreign direct investment (FDI)**. World FDI has grown from $2 trillion in 1990 to $7 trillion in 2000 and $17 trillion in 2009 (Figure 9.7.4).

FDI does not flow equally around the world. Only 30 percent of FDI in 2009 went from a developed to a developing country, whereas 70 percent moved between two developed countries. Among developing regions, more than one-fourth each was directed to East Asia and Latin America (Figure 9.7.5).

The major sources of FDI are transnational corporations (TNCs). A transnational corporation invests and operates in countries other than the one in which its headquarters are located. Of the 100 largest TNCs in 2009, 61 had headquarters in Europe, 19 in the United States, 10 in Japan, 3 in other developed countries, and only 7 in developing countries.

▲ 9.7.3 **FOREIGN DIRECT INVESTMENT**
Japanese carmakers have built several assembly plants in Thailand.

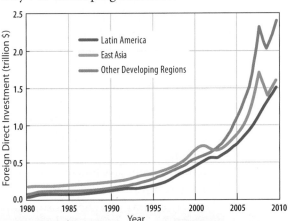

▲ 9.7.4 **GROWTH IN FDI**

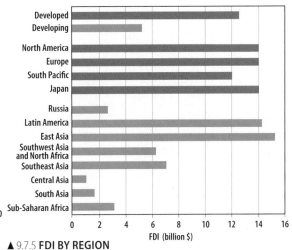

▲ 9.7.5 **FDI BY REGION**

▲ 9.7.6 **FOCUS ON SOUTHEAST ASIA: INTERNATIONAL TRADE**
Child labor in clothing factory.

FOCUS ON SOUTHEAST ASIA

Southeast Asia has become a major manufacturer of textiles and clothing, taking advantage of cheap labor. Thailand has become the region's center for the manufacturing of automobiles and other consumer goods. Indonesia, the world's fourth most populous country, is a major producer of petroleum (Figure 9.7.6). Development has slowed because of painful reforms to restore confidence among international investors shaken by unwise and corrupt investments made possible by lax regulations and excessively close cooperation among manufacturers, financial institutions, and government agencies.

9.8 Financing Development

▶ **Developing countries finance some development through foreign aid and loans.**
▶ **To qualify for loans, a country may need to enact economic reforms.**

Developing countries lack the money needed to finance development. So they obtain grants and loans from governments, banks, and international organizations based in developed countries.

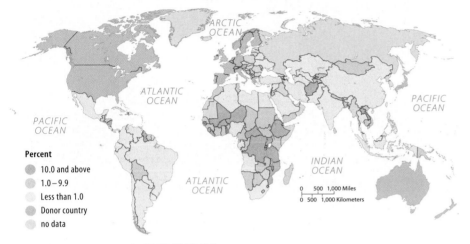

Percent
- 10.0 and above
- 1.0 – 9.9
- Less than 1.0
- Donor country
- no data

▲ 9.8.1 **FOREIGN AID AS PERCENT OF GNI**

- $1 billion and above
- $100 – 999 million
- below $100 million
- developed country (Europe, North America, South Pacific)
- no data

▲ 9.8.2 **DEVELOPMENT ASSISTANCE**

▼ 9.8.3 **WORLD BANK INVESTMENT: THE PHILIPPINES**
Wind farm in Bangui Bay, the Philippines, financed with a World Bank loan.

FOREIGN AID

Most developing countries also receive aid directly from governments of developed countries. The U.S. government allocates approximately 0.2 percent (1/5 of 1%) of its GNI to foreign aid. European countries average a good bit more, approximately 0.5 percent (Figure 9.8.1).

LOANS

The two major international lending organizations are the World Bank and the International Monetary Fund (IMF). The World Bank and IMF were conceived in 1944 to promote development after the devastation of World War II and to avoid a repetition of the disastrous economic policies contributing to the Great Depression of the 1930s. The IMF and World Bank became specialized agencies of the United Nations when it was established in 1945. Twenty-six countries received at least $1 billion in 2009 (Figure 9.8.2).

Developing countries borrow money to build new infrastructure, such as hydroelectric dams, electric transmission lines, flood-protection systems, water supplies, roads, and hotels (Figure 9.8.3). The theory is that the new infrastructure attracts businesses, which in turn pays taxes used to repay the loans and to improve people's living conditions.

In reality, the World Bank itself judges half of the projects it has funded in Africa to be failures. Common reasons include:

- Projects do not function as intended because of faulty engineering.
- Aid is squandered, stolen, or spent on armaments by recipient nations.
- New infrastructure does not attract other investment.

STRUCTURAL ADJUSTMENT PROGRAMS

Some developing countries have had difficulty repaying their loans. The IMF, World Bank, and banks in developed countries fear that granting, canceling, or refinancing debts without strings attached would perpetuate bad habits in developing countries. Therefore before getting debt relief, a developing country is required to prepare a Policy Framework Paper (PFP) outlining a structural adjustment program.

A **structural adjustment program** includes economic "reforms" or "adjustments." Requirements placed on a developing country typically include:

- Spend only what it can afford.
- Direct benefits to the poor not just the elite.
- Divert investment from military to health and education spending.
- Invest scarce resources where they would have the most impact.
- Encourage a more productive private sector.
- Reform the government, including a more efficient civil service, more accountable fiscal management, more predictable rules and regulations, and more dissemination of information to the public.

Critics charge that poverty worsens under structural adjustment programs. By placing priority on reducing government spending and inflation, structural adjustment programs may result in:

- Cuts in health, education, and social services that benefit the poor.
- Higher unemployment.
- Loss of jobs in state enterprises and the civil service.
- Less support for those most in need, such as poor pregnant women, nursing mothers, young children, and elderly people.

In short, structural reforms allegedly punish Earth's poorest people for actions they did not commit—waste, corruption, misappropriation, military build-ups.

International organizations respond that the poor suffer more when a country does not undertake reforms. Economic growth is what benefits the poor the most in the long run. Nevertheless, in response to criticisms, the IMF

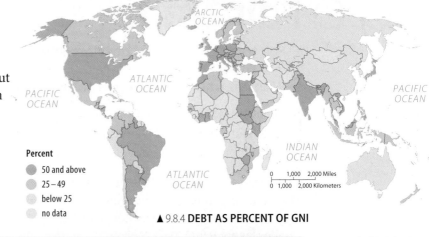

Percent
- 50 and above
- 25 – 49
- below 25
- no data

0   1,000   2,000 Miles
0   1,000   2,000 Kilometers

▲ 9.8.4 **DEBT AS PERCENT OF GNI**

◄ 9.8.5 **WORLD BANK INVESTMENT IN AFGHANISTAN**
Use Google Earth™ to explore development aid in Afghanistan.

Fly to: *Kabul Airport, Afghanistan.*
Drag to: *Enter street view.*
Exit: *Ground level view.*
Zoom out until airport is visible.

The World Bank paid for the long straight dark narrow strip.

1. What is it?
2. Does it appear to be in good condition or in poor condition?

and World Bank now encourage innovative programs to reduce poverty and corruption, and consult more with average citizens. A safety net must be included to ease short-term pain experienced by poor people. Meanwhile, in the twenty-first century, it is the developed countries that have piled up the most debt, especially in the wake of the severe recession of 2007-09 (Figure 9.8.4).

FOCUS ON CENTRAL ASIA

Within Central Asia, the level of development is relatively high in Kazakhstan and Iran. Not by coincidence, these two countries are the region's leading producers of petroleum. In Kazakhstan, rising oil revenues are being used to finance a carefully managed improvement in overall development. In Iran, a large share of the rising oil revenues has been used to maintain low consumer prices rather than to promote development.

Since coming to power in a 1979 revolution, Iran's Shiite leaders have also used oil revenues to promote revolutions elsewhere in the region and to sweep away elements of development and social customs they perceive to be influenced by Europe or North America. War-torn Afghanistan has received more development assistance than any other country in recent years (Figure 9.8.5).

9.9 Fair Trade

► **Fair trade is a model of development that is meant to protect small businesses and workers.**

► **With fair trade, a higher percentage of the sales price goes back to the producers.**

A variation of the international trade model of development is **fair trade**, in which products are made and traded following practices and standards that protect workers and small businesses in developing countries.

Two sets of standards distinguish fair trade:

• Fairtrade Labelling Organizations International (FLO) sets international standards for fair trade (Figure 9.9.1).

• Standards applied to workers on farms and in factories.

► 9.9.1 **FAIR TRADE CLOTHING LABEL** Fair trade label in shirt.

FAIR TRADE PRODUCER PRACTICES

Many farmers and artisans in developing countries are unable to borrow from banks the money they need to invest in their businesses. By banding together in fair trade cooperatives, they can get credit, reduce their raw material costs, and maintain higher and fairer prices for their products (Figure 9.9.2).

Cooperatives are managed democratically, so farmers and artisans learn leadership and organizational skills. The people who grow or make the products have a say in how local resources are utilized and sold. Safe and healthy working conditions can be protected. Cooperatives thus benefit the local farmers and artisans who are members, rather than absentee corporate owners interested only in maximizing profits.

For fair trade coffee, consumers pay prices comparable to those charged by gourmet brands. However, fair trade coffee producers receive a significantly higher price per pound than traditional coffee producers: around $1.20 compared to around $0.80 per pound. Through bypassing exploitative middlemen and working directly with producers, fair trade organizations are able to cut costs and return a greater percentage of the retail price to the producers.

In North America, fair trade products have been primarily craft products such as decorative home accessories, jewelry, textiles, and ceramics. Ten Thousand Villages is the largest fair trade organization in North America specializing in handicrafts. In Europe, most fair trade sales are in food, including coffee, tea, bananas, chocolate, cocoa, juice, sugar, and honey products. TransFair USA certifies the products sold in the United States that are fair trade.

◄ 9.9.2 **FAIR TRADE FOOD**
Fair trade rice for export in Dehradun, India.

The gap between developed and developing countries is narrowing in health and education. For example, during the 1950s people lived on average more than two decades longer in developed countries than in developing ones. In the twenty-first century, the gap is less than ten years (Figure 9.10.4). On the other hand, the gap in wealth between developed and developing countries has widened (Figure 9.10.5).

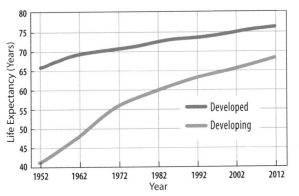

▲ 9.10.4 **CHANGE IN LIFE EXPECTANCY**

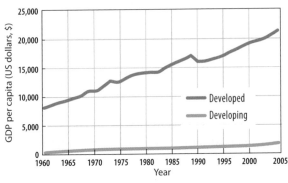

▲ 9.10.5 **CHANGE IN GDP PER CAPITA**

▼ 9.10.6 **FOCUS ON SUB-SAHARAN AFRICA: EDUCATION**
School in Kenya.

EIGHT GOALS

To reduce disparities between developed and developing countries, the United Nations has set eight Millennium Development Goals:

Goal 1: End poverty and hunger
Progress: Extreme poverty has been cut substantially in the world, primarily because of success in Asia, but it has not declined in sub-Saharan Africa.

Goal 2: Achieve universal primary (elementary school) education
Progress: The percentage of children not enrolled in school remains relatively high in South Asia and sub-Saharan Africa.

Goal 3: Promote gender equality and empower women
Progress: Gender disparities remain in all regions, as discussed in Section 9.5.

Goal 4: Reduce child mortality
Progress: Infant mortality rates have declined in most regions, except sub-Saharan Africa.

Goal 5: Improve maternal health
Progress: One-half million women die from complications during pregnancy; 99 percent of these women live in developing countries.

Goal 6: Combat HIV/AIDS, malaria, and other diseases
Progress: The number of people living with HIV remains high, especially in sub-Saharan Africa, as discussed in Chapter 5.

Goal 7: Ensure environmental sustainability
Progress: Water scarcity and quality, deforestation, and overfishing are still especially critical environmental issues, according to the United Nations.

Goal 8: Develop a global partnership for development
Progress: Aid from developed to developing countries has instead been declining.

FOCUS ON SUB-SAHARAN AFRICA

Sub-Saharan Africa has the least favorable prospect for development. The region has the world's highest percentage of people living in poverty and suffering from poor health and low education levels (Figures 9.10.6 and 9.10.7). And conditions are getting worse: the average African consumes less today than a quarter-century ago. The fundamental problem in many countries of sub-Saharan Africa is a dramatic imbalance between the number of inhabitants and the capacity of the land to feed the population.

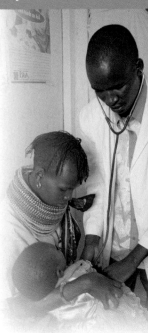

▲ 9.10.7 **FOCUS ON SUB-SAHARAN AFRICA: HEALTH**
Clinic in Kenya.

The world is divided into developed countries and developing ones. Developed and developing countries can be compared according to a number of indicators.

Key Questions

How does development vary among regions?

▶ The United Nations has created the Human Development Index to measure the level of development of every country.

▶ Gross National Income measures the standard of living in a country.

▶ Developed countries display higher levels of education and literacy.

▶ People in developed countries have a longer life expectancy.

▶ A Gender Inequality Index compares the level of development of women and men in every country.

How can countries promote development?

▶ The two principal paths to development are self-sufficiency and international trade.

▶ Self-sufficiency was the most commonly used path in the past, but most countries now follow international trade.

▶ Developing countries finance trade through loans, but may required to undertake economic reforms.

What are future challenges for development?

▶ Fair trade is an alternative approach to development through trade that provides greater benefits to the producers in developing countries.

▶ The United Nations has set Millennium Development Goals for countries to enhance their level of development.

Thinking Geographically

Review the major economic, social, and demographic characteristics that contribute to a country's level of development.

1. Which indicators can vary significantly by gender within countries and between countries at various levels of development? Why?

Some geographers have been attracted to the concepts of Immanuel Wallerstein, who argued that the modern world consists of a single entity, the capitalist world economy that is divided into three regions: the core, semi-periphery, and periphery (refer to Figure 9.10.1).

2. How have the boundaries among these three regions changed?

Opposition to international trade , as well as the severe recession of the early twenty-first century, has encouraged some countries to switch from international trade back to self-sufficiency (Figure 9.CR.1).

3. What are the advantages and challenges of returning to self-sufficiency in poor economic conditions?

On the Internet

Each year's United Nations Human Development Index Report, including numerous indicators for every country, can be accessed at **http://hdr.undp.org,** or scan QR on first page of this chapter.

Indicators cited in this chapter that are not part of the HDI can be found through the Earth Trends portion of the World Resources Institute (WRI) web site at **http://earthtrends.wri.org/.**

Several data sources, including the United Nations and the CIA, are brought together at **www.NationMaster.com.**

◀ 9.CR.1 **WTO PROTESTS**
Protesters at WTO meeting in Seattle, 1999.

Interactive Mapping

INTERNAL VARIATIONS IN DEVELOPMENT

The level of development varies within Latin America's two most populous countries, Brazil and Mexico.

Open MapMaster Latin America in Mastering**GEOGRAPHY**™

Select: *Economic* then *Mapping Poverty and Prosperity.*
Select: *Population* then *Population Density.*

Use the slider tool to adjust the layer opacity.

Are high population concentrations within Brazil and Mexico found primarily in the poorer or the wealthier regions of the two countries?

Explore

BRASILIA

A number of countries have built or are considering constructing new cities to promote development in poorer regions. One example is Brasilia, which was started in the 1950s and became the capital of Brazil in 1960.

Fly to: *National Congress of Brazil, Brasilia, Brazil*
Click on box in middle of screen.

What is the predominant style of housing constructed for the residents?

Key Terms

Adolescent fertility rate
The number of births per 1,000 women age 15-19.

Developed country (more developed country or MDC)
A country that has progressed relatively far along a continuum of development.

Developing country (less developed country or LDC)
A country that is at a relatively early stage in the process of economic development.

Development
A process of improvement in the material conditions of people through diffusion of knowledge and technology.

Fair trade
Alternative to international trade that emphasizes small businesses and worker-owned and democratically run cooperatives and requires employers to pay workers fair wages, permit union organizing, and comply with minimum environmental and safety standards.

Foreign direct investment
Investment made by a foreign company in the economy of another country.

Gender Inequality Index (GII)
Indicator constructed by the United Nations to measure the extent of each country's gender inequality.

Gross domestic product (GDP)
The value of the total output of goods and services produced in a country in a year, not accounting for money that leaves and enters the country.

Gross national income (GNI)
The value of the output of goods and services produced in a country in a year, including money that leaves and enters the country.

Human Development Index (HDI)
Indicator of level of development for each country, constructed by United Nations, combining income, literacy, education, and life expectancy.

Inequality-adjusted HDI (IHDI)
Indicator of level of development for each country that modifies the HDI to account for inequality.

Literacy rate
The percentage of a country's people who can read and write.

Maternal mortality ratio
The number of women who die giving birth per 100,000 births.

Primary sector
The portion of the economy concerned with the direct extraction of materials from Earth's surface, generally through agriculture, although sometimes by mining, fishing, and forestry.

Productivity
The value of a particular product compared to the amount of labor needed to make it.

Secondary sector
The portion of the economy concerned with manufacturing useful products through processing, transforming, and assembling raw materials.

Structural adjustment program
Economic policies imposed on less developed countries by international agencies to create conditions encouraging international trade, such as raising taxes, reducing government spending, controlling inflation, selling publicly owned utilities to private corporations, and charging citizens more for services.

Tertiary sector
The portion of the economy concerned with transportation, communications, and utilities, sometimes extended to the provision of all goods and services to people in exchange for payment.

Value added
The gross value of the product minus the costs of raw materials and energy.

► **LOOKING AHEAD**

One of the most fundamental differences between developed and developing countries are the predominant methods of agriculture.

10 Food and Agriculture

When you buy food in the supermarket, are you reminded of a farm? Not likely. The meat is carved into pieces that no longer resemble an animal and is wrapped in paper or plastic film. Often the vegetables are canned or frozen. The milk and eggs are in cartons.

Providing food in the United States and Canada is a vast industry. Only a few people are full-time farmers, and they may be more familiar with the operation of computers and advanced machinery than the typical factory or office worker is.

The mechanized, highly productive American or Canadian farm contrasts with the subsistence farm found in much of the world. In China and India, more than half of the people are farmers who grow enough food for themselves and their families to survive, with little surplus. This sharp contrast in agricultural practices constitutes one of the most fundamental differences between the world's developed countries and developing countries.

What do people eat?

10.1 **Origin of Agriculture**

10.2 **Diet**

10.3 **Nutrition and Hunger**

ORGANIC FARM,
KASRAWAD, INDIA

No.- 27
ORGANIC

How is agriculture distributed?

What challenges does agriculture face?

IR64680-81-2-2-1-3
BASAL   24 DAT   PI
45   45   45
Kg N / ha

10.1 Origin of Agriculture

▶ **Early humans obtained food through hunting and gathering.**

▶ **Agriculture originated in multiple hearths and diffused in many directions.**

Agriculture is the deliberate modification of Earth's surface through cultivation of plants and rearing of animals to obtain sustenance or economic gain. Agriculture originated when humans domesticated plants and animals for their use. The word cultivate means "to care for," and a **crop** is any plant cultivated by people.

HUNTERS AND GATHERERS

Before the invention of agriculture, all humans probably obtained the food they needed for survival through hunting for animals, fishing, or gathering plants (including berries, nuts, fruits, and roots). Hunters and gatherers lived in small groups, with usually fewer than 50 persons, because a larger number would quickly exhaust the available resources within walking distance.

Typically, the men hunted game or fished, and the women collected berries, nuts, and roots. This division of labor sounds like a stereotype but is based on evidence from archaeology and anthropology, although exceptions to this pattern have been documented. They collected food often, perhaps daily. The food search might take only a short time or much of the day, depending on local conditions.

The group traveled frequently, establishing new home bases or camps. The direction and frequency of migration depended on the movement of game and the seasonal growth of plants at various locations. We can assume that groups communicated with each other concerning hunting rights, intermarriage, and other specific subjects. For the most part, they kept the peace by steering clear of each other's territory.

Today perhaps a quarter-million people still survive by hunting and gathering rather than by agriculture. Examples include the Spinifex (also known as Pila Nguru) people, who live in Australia's Great Victorian Desert; the Sentinelese people, who live in India's Andaman Islands; and the Bushmen, who live in Botswana and Namibia (Figure 10.1.1). Contemporary hunting and gathering societies are isolated groups living on the periphery of world settlement, but they provide insight into human customs that prevailed in prehistoric times, before the invention of agriculture.

▲ 10.1.1 **HUNTERS AND GATHERERS** Botswana.

CROP HEARTHS

Why did most nomadic groups convert from hunting, gathering, and fishing to agriculture? Geographers and other scientists agree that agriculture originated in multiple hearths around the world. They do not agree on when agriculture originated and diffused, or why. Early centers of crop domestication include Southwest Asia, sub-Saharan Africa, Latin America, East Asia, and Southeast Asia (Figure 10.1.2). Crop cultivation diffused from these multiple hearths:

- From Southwest Asia: west to Europe and east to Central Asia.

- From sub-Saharan Africa: south to southern Africa.

- From Latin America: north to North America and south to tropical South America.

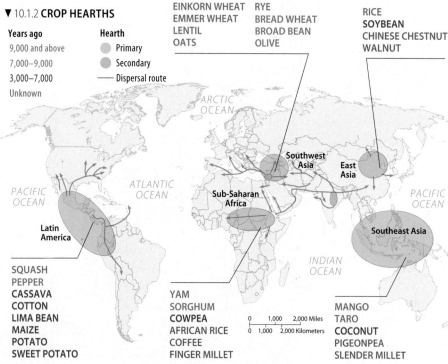

▼ 10.1.2 **CROP HEARTHS**

Years ago
9,000 and above
7,000–9,000
3,000–7,000
Unknown

Hearth
Primary
Secondary
— Dispersal route

BARLEY
EINKORN WHEAT
EMMER WHEAT
LENTIL
OATS

RYE
BREAD WHEAT
BROAD BEAN
OLIVE

RICE
SOYBEAN
CHINESE CHESTNUT
WALNUT

ARCTIC OCEAN

Southwest Asia

East Asia

Sub-Saharan Africa

PACIFIC OCEAN

ATLANTIC OCEAN

PACIFIC OCEAN

Latin America

Southeast Asia

INDIAN OCEAN

SQUASH
PEPPER
CASSAVA
COTTON
LIMA BEAN
MAIZE
POTATO
SWEET POTATO

YAM
SORGHUM
COWPEA
AFRICAN RICE
COFFEE
FINGER MILLET

0   1,000   2,000 Miles
0   1,000   2,000 Kilometers

MANGO
TARO
COCONUT
PIGEONPEA
SLENDER MILLET

ANIMAL HEARTHS

Animals were also domesticated in multiple hearths at various dates. Southwest Asia is thought to be the hearth for the domestication of the largest number of animals that would prove to be most important for agriculture. Animals thought to be domesticated in Southwest Asia between 8,000 and 9,000 years ago include cattle, goats, pigs, and sheep. (Figure 10.1.3). The turkey is thought to have been domesticated in the Western Hemisphere (Figure 10.1.4).

Inhabitants of Southwest Asia may have been the first to integrate cultivation of crops with domestication of herd animals such as cattle, sheep, and goats. These animals were used to plow the land before planting seeds and, in turn, were fed part of the harvested crop. Other animal products, such as milk, meat, and skins, may have been exploited at a later date. This integration of plants and animals is a fundamental element of modern agriculture.

Domestication of the dog is thought to date from around 12,000 years ago, also in Southwest Asia. The horse is considered to have been domesticated in Central Asia; diffusion of the domesticated horse may have been associated with the diffusion of the Indo-European language, as discussed in Chapter 5.

▲ 10.1.3 **ANIMAL HEARTHS**

Years ago
12,000
9,000
8,000
6,000
Unknown

WHY AGRICULTURE ORIGINATED

Scientists do not agree on whether agriculture originated primarily because of environmental factors or cultural factors. Probably a combination of both factors contributed.

- **Environmental Factors.** The first domestication of crops and animals around 10,000 years ago coincides with climate change. This marked the end of the last ice age, when permanent ice cover receded from Earth's midlatitudes to polar regions, resulting in a massive redistribution of humans, other animals, and plants at that time.

- **Cultural Factors.** Preference for living in a fixed place rather than as nomads may have led hunters and gatherers to build permanent settlements and to store surplus vegetation there.

In gathering wild vegetation, people inevitably cut plants and dropped berries, fruits, and seeds. These hunters probably observed that, over time, damaged or discarded food produced new plants. They may have deliberately cut plants or dropped berries on the ground to see if they would produce new plants.

Subsequent generations learned to pour water over the site and to introduce manure and other soil improvements. Over thousands of years, plant cultivation apparently evolved from a combination of accident and deliberate experiment.

That agriculture had multiple origins means that, from earliest times, people have produced food in distinctive ways in different regions. This diversity derives from a unique legacy of wild plants, climatic conditions, and cultural preferences in each region.

▼ 10.1.4 **WILD TURKEY** California.

10.2 Diet

► **Most people derive most of their food energy from cereals.**
► **Climate and the level of development influence choice of food.**

Everyone needs food to survive. Consumption of food varies around the world, both in total amount and source of nutrients. The variation results from a combination of:

- **Level of development.** People in developed countries tend to consume more food and from different sources than do people in developing countries.
- **Physical conditions.** Climate is important in influencing what can be most easily grown and therefore consumed in developing countries. In developed countries, though, food is shipped long distances to locations with different climates.
- **Cultural preferences.** Some food preferences and avoidances are expressed without regard for physical and economic factors, as discussed in Chapter 4.

TOTAL CONSUMPTION OF FOOD

Dietary energy consumption is the amount of food that an individual consumes. The unit of measurement of dietary energy is a kilocalorie (kcal), or Calorie in the United States. One gram (or ounce) of each food source delivers a kcal level that nutritionists can measure.

Humans derive most of their kilocalories through consumption of **cereal grain**, or simply cereal, which is a grass that yields grain for food. The three leading cereal grains—maize (corn in North America), wheat, and rice—together account for nearly 90 percent of all grain production and more than 40 percent of all dietary energy consumed worldwide.

Wheat is the principal cereal grain consumed in the developed regions of Europe and North America (Figure 10.2.1). Wheat is consumed in the form of bread, pasta, cake, and many

▲ 10.2.2 **WHEAT CONSUMPTION**
Lille, France.

other forms (Figure 10.2.2). It is also the most consumed **grain** in the developing regions of Central and Southwest Asia, where relatively dry conditions are more suitable for growing wheat than other grains.

Rice is the principal cereal grain consumed in the developing regions of East, South, and Southeast Asia (Figure 10.2.3). It is the most suitable cereal crop for production in tropical climates.

▼ 10.2.3 **RICE CONSUMPTION**
Ho Chi Minh City, Vietnam.

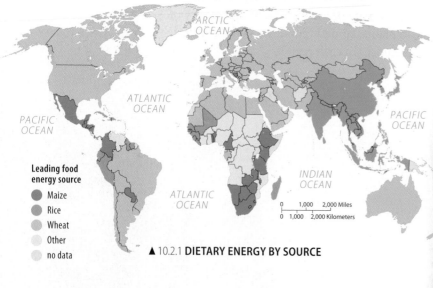

Leading food energy source
- Maize
- Rice
- Wheat
- Other
- no data

▲ 10.2.1 **DIETARY ENERGY BY SOURCE**

Maize is the leading crop in the world, though much of it is grown for purposes other than direct human consumption, especially as animal feed. It is the leading crop in some countries of sub-Saharan Africa (Figure 10.2.4).

A handful of countries obtain the largest share of dietary energy from other crops, especially in sub-Saharan Africa. These include cassava, sorghum, millet, plantains, sweet potatoes, and yams (Figure 10.2.5). Sugar is the leading source of dietary energy in several Latin American countries.

▲ 10.2.5 **YAM CONSUMPTION**
Ghana.

▲ 10.2.4 **MAIZE CONSUMPTION**
Weekly market, Madagascar.

SOURCE OF NUTRIENTS

Protein is a nutrient needed for growth and maintenance of the human body. Many food sources provide protein of varying quantity and quality. One of the most fundamental differences between developed and developing regions is the primary source of protein (Figure 10.2.6). In developed countries, the leading source of protein is meat products, including beef, pork, and poultry (Figure 10.2.7). Meat accounts for approximately one-third of all protein intake in developed countries, compared to approximately one-tenth in developing ones (Figure 10.2.8). In most developing countries, cereal grains provide the largest share of protein.

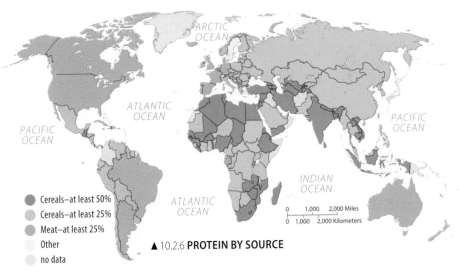

- Cereals—at least 50%
- Cereals—at least 25%
- Meat—at least 25%
- Other
- no data

▲ 10.2.6 **PROTEIN BY SOURCE**

▲ 10.2.7 **MEAT CONSUMPTION**
Dublin, Ireland.

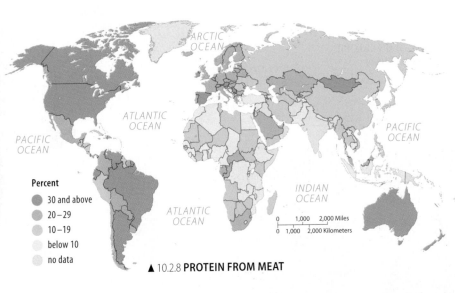

Percent
- 30 and above
- 20–29
- 10–19
- below 10
- no data

▲ 10.2.8 **PROTEIN FROM MEAT**

10.3 Nutrition and Hunger

▶ **On average, the world produces enough food to meet dietary needs.**

▶ **Some developing countries lack food security and are undernourished.**

The United Nations defines **food security** as physical, social, and economic access at all times to safe and nutritious food sufficient to meet dietary needs and food preferences for an active and healthy life. By this definition, roughly one-eighth of the world's inhabitants do not have food security.

DIETARY ENERGY NEEDS

To maintain a moderate level of physical activity, according to the United Nations Food and Agricultural Organization, an average individual needs to consume on a daily basis at least 1,800 kcal.

Average consumption worldwide is 2,780 kcal per day, or roughly 50 percent more than the recommended minimum. Thus, most people get enough food to survive. People in developed countries are consuming on average nearly twice the recommended minimum, 3,470 kcal per day (Figure 10.3.1). The United States has the world's highest consumption, 3,800 kcal per day per person. The consumption of so much food is one reason that obesity rather than hunger is more prevalent in the United States, as well as other developed countries (Figure 10.3.2).

In developing regions, average daily consumption is 2,630 kcal, still above the recommended minimum. However, the average in sub-Saharan Africa is only 2,290, an indication that a large percentage of Africans are not getting enough to eat. Diets are more likely to be deficient in countries where people have to spend a high percentage of their income to obtain food (Figure 10.3.3).

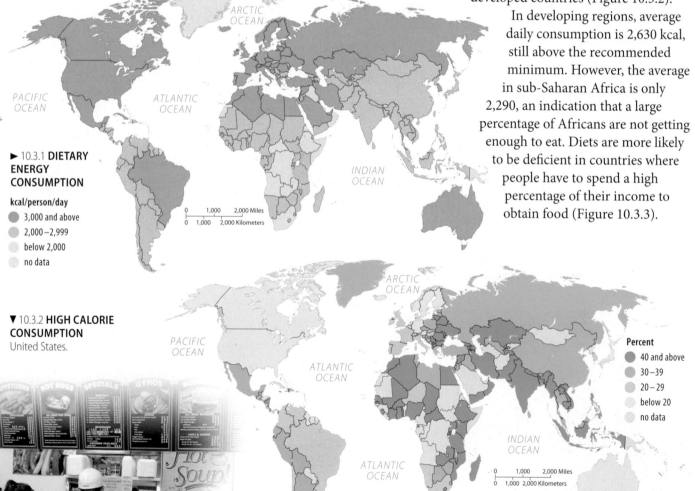

▶ 10.3.1 **DIETARY ENERGY CONSUMPTION**

kcal/person/day
- 3,000 and above
- 2,000–2,999
- below 2,000
- no data

▼ 10.3.2 **HIGH CALORIE CONSUMPTION**
United States.

Percent
- 40 and above
- 30–39
- 20–29
- below 20
- no data

▲ 10.3.3 **INCOME SPENT ON FOOD**

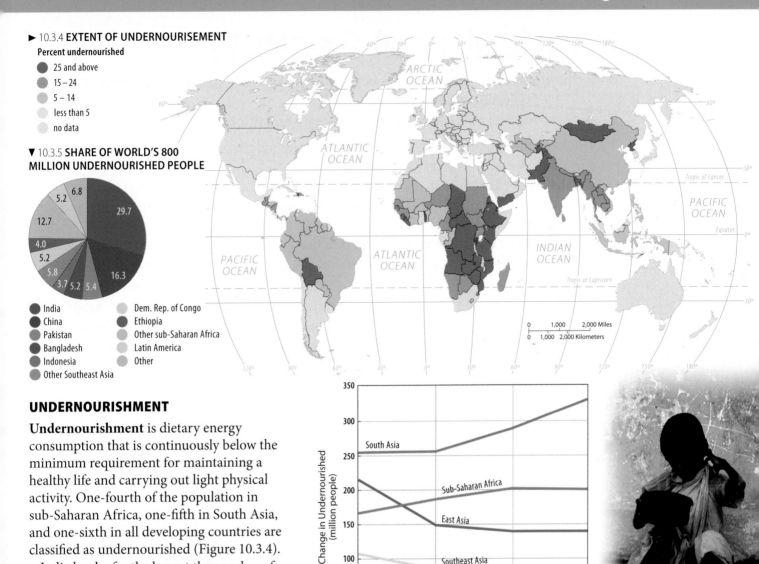

► 10.3.4 **EXTENT OF UNDERNOURISEMENT**

Percent undernourished

- 25 and above
- 15 – 24
- 5 – 14
- less than 5
- no data

▼ 10.3.5 **SHARE OF WORLD'S 800 MILLION UNDERNOURISHED PEOPLE**

- India — 29.7
- China — 16.3
- Pakistan — 5.4
- Bangladesh — 5.2
- Indonesia — 3.7
- Other Southeast Asia — 5.8
- Dem. Rep. of Congo — 5.2
- Ethiopia — 4.0
- Other sub-Saharan Africa — 12.7
- Latin America — 5.2
- Other — 6.8

UNDERNOURISHMENT

Undernourishment is dietary energy consumption that is continuously below the minimum requirement for maintaining a healthy life and carrying out light physical activity. One-fourth of the population in sub-Saharan Africa, one-fifth in South Asia, and one-sixth in all developing countries are classified as undernourished (Figure 10.3.4).

India has by far the largest the number of undernourished people, 238 million, followed by China with 130 million (Figure 10.3.5). Worldwide, the total number of undernourished people has not changed much in several decades (Figure 10.3.6).

▲ 10.3.6 **CHANGE IN NUMBER UNDERNOURISHED**

(Graph: Change in Undernourished (million people) vs Year 1991–2006, showing lines for South Asia, Sub-Saharan Africa, East Asia, Southeast Asia, Latin America, Southwest Asia & North Africa, Developed Regions)

▲ 10.3.7
UNDERNOURISHMENT
Somalia.

AFRICA'S FOOD-SUPPLY STRUGGLE

Sub-Saharan Africa is struggling to keep food production ahead of population growth (Figure 10.3.7). Since 1961, food production has increased substantially in sub-Saharan Africa, but so has population (Figure 10.3.8). As a result, food production per capita has changed little in a half-century.

The threat of famine is particularly severe in the Sahel. Traditionally, this region supported limited agriculture. With rapid population growth, farmers overplanted, and herd size increased beyond the capacity of the land to support the animals. Animals overgrazed the limited vegetation and clustered at scarce water sources.

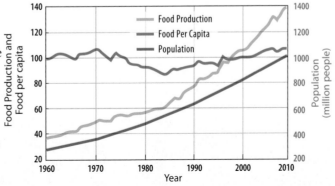

◄ 10.3.8 **POPULATION AND FOOD PRODUCTION IN AFRICA**

(Graph: Food Production and Food per capita / Population (million people) vs Year 1960–2010, with lines for Food Production, Food Per Capita, and Population)

10.4 Agricultural Regions

▶ **The world can be divided into several regions of subsistence agriculture and commercial agriculture.**

▶ **These regions are related in part to climate conditions.**

The most fundamental differences in agricultural practices are between subsistence agriculture and commercial agriculture.

- **Subsistence agriculture** is generally practiced in developing countries (Figure 10.4.1). It is designed primarily to provide food for direct consumption by the farmer and the farmer's family (Figure 10.4.2).

- **Commercial agriculture,** generally practiced in developed countries, is undertaken primarily to generate products for sale off the farm to food-processing companies (Figure 10.4.3).

The most widely used map of world agricultural regions was prepared by geographer Derwent Whittlesey in 1936. Climate regions played an important role in determining agricultural regions, such as pastoral nomadism (Figure 10.4.4) and mixed crop and livestock (Figure 10.4.5).

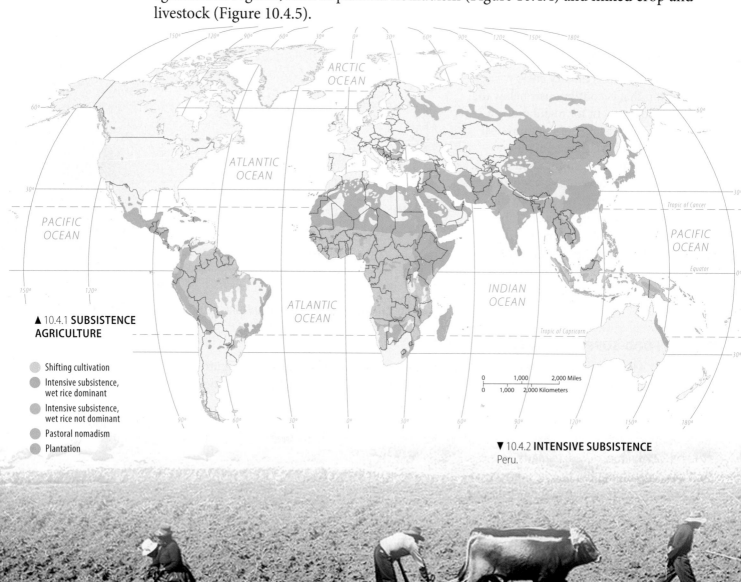

▲ 10.4.1 **SUBSISTENCE AGRICULTURE**

- ⬤ Shifting cultivation
- ⬤ Intensive subsistence, wet rice dominant
- ⬤ Intensive subsistence, wet rice not dominant
- ⬤ Pastoral nomadism
- ⬤ Plantation

▼ 10.4.2 **INTENSIVE SUBSISTENCE**
Peru.

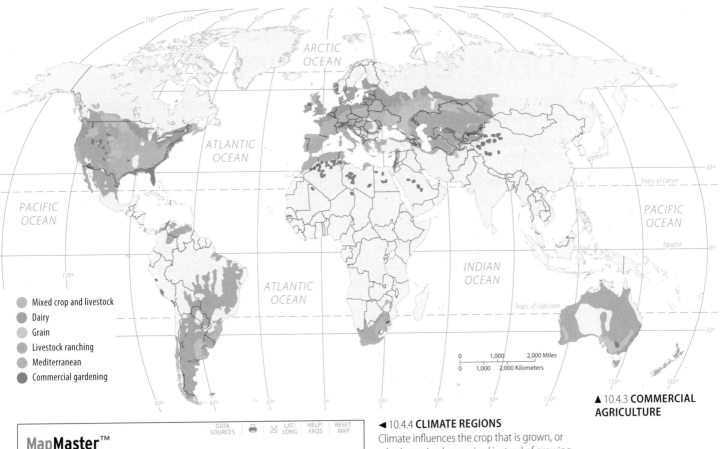

- Mixed crop and livestock
- Dairy
- Grain
- Livestock ranching
- Mediterranean
- Commercial gardening

▲ 10.4.3 **COMMERCIAL AGRICULTURE**

◄ 10.4.4 **CLIMATE REGIONS**
Climate influences the crop that is grown, or whether animals are raised instead of growing any crop.

Launch MapMaster Southwest Asia and North Africa in Mastering**GEOGRAPHY**™

Select: *Climate* from the *Physical Environment* menu, then *Agricultural Regions* from the *Economic* menu.

What climate region is correlated with pastoral nomadism?

MapMaster™

| DATA SOURCES | | LAT/ LONG | HELP/ FAQS | RESET MAP |

A TROPICAL AND HUMID CLIMATES
- Af   Tropical wet climate
- Aw   Tropical savanna climate

B DRY CLIMATES
- BWh   Tropical and subtropical desert
- BSh   Tropical and subtropical steppe
- BSk   Midlatitude steppe

C MILD MIDLATITUDE CLIMATES
- Cs   Mediterranean summer—dry

H HIGHLAND
- H   Complex mountain climates

0   250   500 Miles
0   250   500 Kilometers

▼ 10.4.5 **MIXED CROP AND LIVESTOCK**
France.

10.5 Comparing Subsistence and Commercial Agriculture

▶ **Subsistence farming is characterized by small farms, a high percentage of farmers, and few machines.**

▶ **Commercial farming has large farms, a small percentage of farmers, and many machines.**

Subsistence and commercial farming differ in several key ways.

FARM SIZE

The average farm size is much larger in commercial agriculture. For example, farms average about 161 hectares (418 acres) in the United States, compared to about 1 hectare in China.

Commercial agriculture is dominated by a handful of large farms. In the United States, the largest 5 percent of farms produce 75 percent of the country's total agriculture. Despite their size, most commercial farms in developed countries are family owned and operated—90 percent in the United States. Commercial farmers frequently expand their holdings by renting nearby fields.

Large size is partly a consequence of mechanization, as discussed below. Combines, pickers, and other machinery perform most efficiently at very large scales, and their considerable expense cannot be justified on a small farm. As a result of the large size and the high level of mechanization, commercial agriculture is an expensive business.

Farmers spend hundreds of thousands of dollars to buy or rent land and machinery before beginning operations. This money is frequently borrowed from a bank and repaid after the output is sold.

The United States had 13 percent more farmland in 2000 than in 1900, primarily through irrigation and reclamation. However, in the twenty-first century it has been losing 1.2 million hectares (3 million acres) per year of its 400 million hectares (1 billion acres) of farmland, primarily because of expansion of urban areas.

PERCENTAGE OF FARMERS IN SOCIETY

In developed countries around 5 percent of workers are engaged directly in farming, compared to around 50 percent in developing countries (Figure 10.5.1). The percentage of farmers is even lower in North America, only around 2 percent. Yet the small percentage of farmers in the United States and Canada produces enough food not only for themselves and the rest of the region but also a surplus to feed people elsewhere.

The number of farmers declined dramatically in developed countries during the twentieth century. The United States had 60 percent fewer farms and 85 percent fewer farmers in 2000 than in 1900. The number of farms in the United States declined from about 6 million farms in 1940 to 4 million in 1960 and 2 million in 1980. Both push and pull migration factors have been responsible for the decline: people were pushed away from farms by lack of opportunity to earn a decent income, and at the same time they were pulled to higher-paying jobs in urban areas. The number of U.S. farmers has stabilized since 1980 at around 2 million.

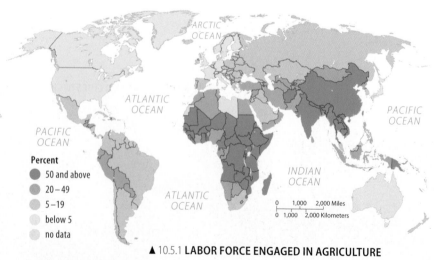

Percent
- 50 and above
- 20–49
- 5–19
- below 5
- no data

▲ 10.5.1 **LABOR FORCE ENGAGED IN AGRICULTURE**

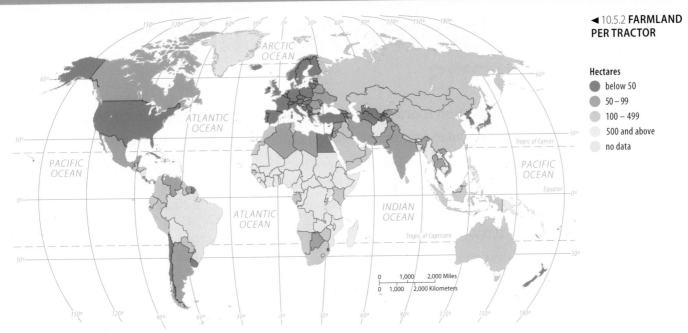

◄ 10.5.2 **FARMLAND PER TRACTOR**

Hectares
- below 50
- 50 – 99
- 100 – 499
- 500 and above
- no data

USE OF MACHINERY

In developed countries, a small number of farmers can feed many people because they rely on machinery to perform work, rather than relying on people or animals (Figure 10.5.2). In developing countries, farmers do much of the work with hand tools and animal power.

Traditionally, the farmer or local craftspeople made equipment from wood, but beginning in the late eighteenth century, factories produced farm machinery. The first all-iron plow was made in the 1770s and was followed in the nineteenth and twentieth centuries by inventions that made farming less dependent on human or animal power. Tractors, combines, corn pickers, planters, and other factory-made farm machines have replaced or supplemented manual labor (Figure 10.5.3).

Transportation improvements also aid commercial farmers. The building of railroads in the nineteenth century, and highways and trucks in the twentieth century, have enabled farmers to transport crops and livestock farther and faster. Cattle arrive at market heavier and in better condition when transported by truck or train than when driven on hoof. Crops reach markets without spoiling.

Commercial farmers use scientific advances to increase productivity. Experiments conducted in university laboratories, industry, and research organizations generate new fertilizers, herbicides, hybrid plants, animal breeds, and farming practices, which produce higher crop yields and healthier animals. Access to other scientific information has enabled farmers to make more intelligent decisions concerning proper agricultural practices. Some farmers conduct their own on-farm research.

Electronics also aid commercial farmers. Global positioning system (GPS) units determine the precise coordinates for spreading different types and amounts of fertilizers. On large ranches, GPS is also used to monitor the location of cattle. Satellite imagery monitors crop progress. Yield monitors attached to combines determine the precise number of bushels being harvested.

▼ 10.5.3 **USE OF MACHINERY**
Combines harvest wheat, Colorado.

10.6 Subsistence Agriculture Regions

► **Shifting cultivation is practiced in wet lands and pastoral nomadism in dry lands.**

► **Asia's large population concentrations practice intensive subsistence agriculture.**

Three types of subsistence agriculture predominate in developing countries: shifting cultivation, pastoral nomadism, and intensive subsistence. Plantation, a form of commercial agriculture found in developing countries, is also discussed here.

SHIFTING CULTIVATION

Shifting cultivation is practiced in much of the world's humid tropics, which have relatively high temperatures and abundant rainfall. Each year villagers designate for planting an area near the settlement. Before planting, they remove the dense vegetation using axes and machetes.

On a windless day the debris is burned under carefully controlled conditions; consequently,

▼ 10.6.1 **SHIFTING CULTIVATION**
Venezuela.

shifting cultivation is sometimes called **slash-and-burn agriculture** (Figure 10.6.1). The rains wash the fresh ashes into the soil, providing needed nutrients.

The cleared area is known by a variety of names, including *swidden, ladang, milpa, chena,* and *kaingin.* The **swidden** can support crops only briefly, usually 3 years or less, before soil nutrients are depleted. Villagers then identify a new site and begin clearing it, leaving the old swidden uncropped for many years, so that it is again overrun by natural vegetation.

Shifting cultivation is being replaced by logging, cattle ranching, and cultivation of cash crops. Selling timber to builders or raising beef cattle for fast-food restaurants is a more effective development strategy than maintaining shifting cultivation. Defenders of shifting cultivation consider it a more environmentally sound approach for tropical agriculture.

PASTORAL NOMADISM

Pastoral nomadism is a form of subsistence agriculture based on the herding of domesticated animals. It is adapted to dry climates, where planting crops is impossible. Pastoral nomads live primarily in the large belt of arid and semiarid land that includes most of North Africa and Southwest Asia, and parts of Central Asia. The Bedouins of Saudi Arabia and North Africa and the Maasai of East Africa are examples of nomadic groups (Figure 10.6.2).

▼ 10.6.2 **PASTORAL NOMADISM**
Sahara Desert, Africa.

Pastoral nomads depend primarily on animals rather than crops for survival. The animals provide milk, and their skins and hair are used for clothing and tents. Like other subsistence farmers, though, pastoral nomads consume mostly grain rather than meat. Their animals are usually not slaughtered, although dead ones may be consumed. To nomads, the size of their herd is both an important measure of power and prestige and their main security during adverse environmental conditions.

Only about 15 million people are pastoral nomads, but they sparsely occupy about 20 percent of Earth's land area. Nomads used to be the most powerful inhabitants of the dry lands. Today, national governments control the nomadic population, using force, if necessary. Governments force groups to give up pastoral nomadism because they want the land for other uses.

INTENSIVE SUBSISTENCE

In densely populated East, South, and Southeast Asia, most farmers practice **intensive subsistence agriculture.** Because the agricultural density—the ratio of farmers to arable land—is so high in parts of East and South Asia, families must produce enough food for their survival from a very small area of land.

Most of the work is done by hand or with animals rather than with machines, in part due to abundant labor, but largely from lack of funds to buy equipment.

The intensive agriculture region of Asia can be divided between areas where wet rice dominates and areas where it does not. The term **wet rice** refers to the practice of planting rice on dry land in a nursery and then moving the seedlings to a flooded field to promote growth. Wet rice is most easily grown on flat land, because the plants are submerged in water much of the time.

The pressure of population growth in parts of East Asia has forced expansion of areas under rice cultivation (Figure 10.6.3). One method of developing additional land suitable for growing rice is to terrace the hillsides of river valleys (Figure 10.6.4).

▲ 10.6 .3 **RICE PRODUCTION**

Million metric tons
- 100.0 and above
- 10.0 – 99.9
- 1.0 – 9.9
- below 1.0
- no data

▲ 10.6.4 **INTENSIVE SUBSISTENCE FARMING**
Use Google Earth to explore rice farming in Southeast Asia.

Fly to: *Banaue, Philippines.*

Use the mouse to zoom in near the Banaue label until you see a series of brown swirling stripes.

Drag to: *street view on one of the brown swirling stripes.*

Exit ground level view.

1. Is the topography of this region flat or hilly?
2. What are the brown swirling stripes?

PLANTATION AGRICULTURE

A **plantation** is a form of commercial agriculture in developing regions that specializes in one or two crops. They are found primarily in the tropics and subtropics, especially in Latin America, sub-Saharan Africa, and Asia (Figure 10.6.5).

Although situated in developing countries, plantations are often owned or operated by Europeans or North Americans and grow crops for sale primarily in developed countries. Among the most important crops grown on plantations are cotton, sugarcane, coffee, rubber, and tobacco.

Until the Civil War, plantations were important in the U.S. South, where the principal crop was cotton, followed by tobacco and sugarcane. Slaves brought from Africa performed most of the labor until the abolition of slavery and the defeat of the South in the Civil War. Thereafter, plantations declined in the United States; they were subdivided and either sold to individual farmers or worked by tenant farmers.

▼ 10.6.5 **SUGARCANE PLANTATION**
Thailand.

10.7 Commercial Agriculture Regions

▶ **Six main types of commercial agriculture are found in developed countries.**

▶ **The type of agriculture is influenced by physical geography.**

Commercial agriculture in developed countries can be divided into six main types. Each type is predominant in distinctive regions within developed countries, depending largely on climate.

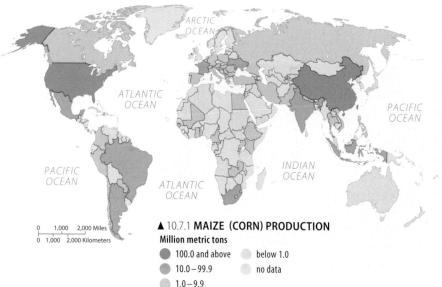

▲ 10.7.1 **MAIZE (CORN) PRODUCTION**

Million metric tons

● 100.0 and above   ○ below 1.0
● 10.0 – 99.9   ○ no data
○ 1.0 – 9.9

MIXED CROP AND LIVESTOCK

The most distinctive characteristic of mixed crop and livestock farming is its integration of crops and livestock. Maize (corn) is the most commonly grown crop (Figure 10.7.1), followed by soybeans. Most of the crops are fed to animals rather than consumed directly by humans. A typical mixed commercial farm devotes nearly all land area to growing crops but derives more than three-fourths of its income from the sale of animal products, such as beef, milk, and eggs.

Mixed crop and livestock farming typically involves **crop rotation.** The farm is divided into a number of fields, and each field is planted on a planned cycle, often of several years duration.

▼ 10.7.2 **DAIRY FARM** Germany.

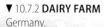

▶ 10.7.3 **MILK PRODUCTION**

Million metric tons

● 10.0 – 100.0   ○ below 1.0
○ 1.0 – 9.9   ○ no data

DAIRY FARMING

Dairy farming is the most important agriculture practiced near large urban areas in developed countries (Figure 10.7.2). Dairy farms must be closer to their markets than other products because milk is highly perishable. The ring surrounding a city from which milk can be supplied without spoiling is known as the **milkshed.**

Traditionally most milk was produced and consumed in developed countries (Figure 10.7.3). However, the share of the world's dairy farming conducted in developing countries has risen dramatically in recent years, and now surpasses the total in developed countries (Figure 10.7.4). Rising incomes permit urban residents to buy more milk products.

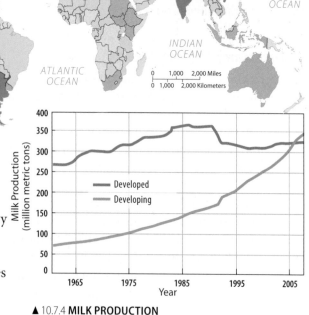

▲ 10.7.4 **MILK PRODUCTION**

GRAIN FARMING

Commercial grain farms are generally located in regions that are too dry for mixed crop and livestock farming, such as the Great Plains of North America (Figure 10.7.5). Unlike mixed crop and livestock farming, crops on a grain farm are grown primarily for consumption by humans rather than by livestock.

The most important crop grown is wheat, used to make flour. It can be stored relatively easily without spoiling and can be transported a long distance. Because wheat has a relatively high value per unit weight, it can be shipped profitably from remote farms to markets.

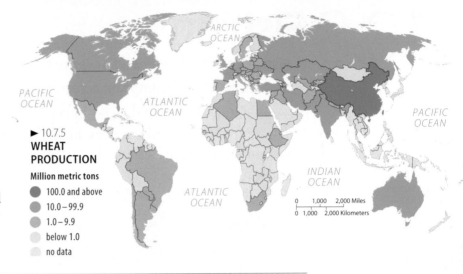

► 10.7.5
WHEAT PRODUCTION
Million metric tons
- 100.0 and above
- 10.0 – 99.9
- 1.0 – 9.9
- below 1.0
- no data

LIVESTOCK RANCHING

Ranching is the commercial grazing of livestock over an extensive area. It is practiced primarily on semiarid or arid land where the vegetation is too sparse and the soil too poor to support crops. China is the leading producer of pig meat, the United States of chicken and beef (Figure 10.7.6).

Ranching has been glamorized in novels and films, although the cattle drives and "Wild West" features of this type of farming actually lasted only a few years in the mid-nineteenth century. Contemporary ranching has become part of the meat-processing industry, rather than carried out on isolated farms.

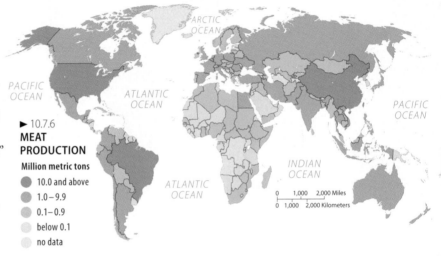

► 10.7.6
MEAT PRODUCTION
Million metric tons
- 10.0 and above
- 1.0 – 9.9
- 0.1 – 0.9
- below 0.1
- no data

COMMERCIAL GARDENING AND FRUIT FARMING

Commercial gardening and fruit farming are the predominant types of agriculture in the U.S. Southeast (Figure 10.7.7). The region has a long growing season and humid climate and is accessible to the large markets in the big cities along the East Coast. It is frequently called **truck farming**, because "truck" was a Middle English word meaning bartering or the exchange of commodities.

Truck farms grow many of the fruits and vegetables that consumers demand in developed countries, such as apples, cherries, lettuce, and tomatoes. A form of truck farming called specialty farming has spread to New England. Farmers are profitably growing crops that have limited but increasing demand among affluent consumers, such as asparagus, mushrooms, peppers, and strawberries.

MEDITERRANEAN AGRICULTURE

Mediterranean agriculture exists primarily on lands that border the Mediterranean Sea and other places that share a similar physical geography, such as California, central Chile, the southwestern part of South Africa, and southwestern Australia. Winters are moist and mild, summers hot and dry. The land is very hilly, and mountains frequently plunge directly to the sea, leaving very little flat land. The two most important crops are olives (primarily for cooking oil) and grapes (primarily for wine).

► 10.7.7 **COMMERCIAL GARDENING**
Peanut farm, Georgia, U.S.A.

10.8 Fishing

▶ **Fish are either caught wild or farmed.**

▶ **Increasing fish consumption is resulting in overfishing.**

The agriculture discussed thus far in this chapter is land-based. Humans also consume food acquired from Earth's waters, including fish, crustaceans (such as shrimp and crabs), molluscs (such as clams and oysters), and aquatic plants (such as watercress).

▶ 10.8.1 **FISH PRODUCTION**

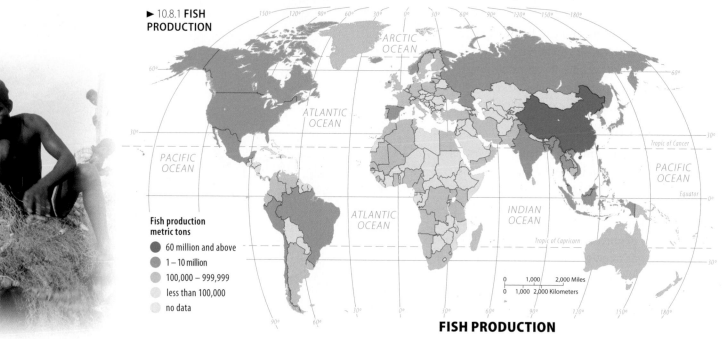

Fish production metric tons
- 60 million and above
- 1 – 10 million
- 100,000 – 999,999
- less than 100,000
- no data

▲ 10.8.2 **FISHING**
Mauritania.

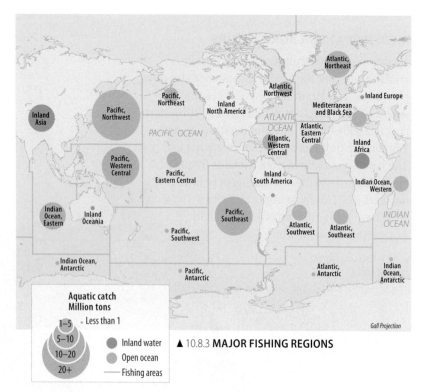

Aquatic catch Million tons
- 1–5   • Less than 1
- 5–10
- 10–20   ● Inland water
- 20+   ● Open ocean
- — Fishing areas

Gall Projection

▲ 10.8.3 **MAJOR FISHING REGIONS**

FISH PRODUCTION

Water-based food is acquired in two ways:

- Fishing, which is the capture of wild fish and other seafood living in the waters.

- **Aquaculture**, or **aquafarming**, which is the cultivation of seafood under controlled conditions.

About two-thirds of the fish caught from the ocean is consumed directly by humans, whereas the remainder is converted to fish meal and fed to poultry and hogs.

China is responsible for 40 percent of the world's yield of fish (Figure 10.8.1). The other leading countries are naturally those with extensive ocean boundaries, including Peru, Indonesia, India, Chile, Japan, and the United States.

The world's oceans are divided into 18 major fishing regions, including seven each in the Atlantic and Pacific oceans and four in the Indian Ocean (Figure 10.8.2). The three areas with the largest yield are all in the Pacific (Figure 10.8.3). Fishing is also conducted in inland waterways, such as lakes and rivers.

FISH CONSUMPTION

At first glance, increased use of food from the sea is attractive. Oceans are vast, covering nearly three-fourths of Earth's surface and lying near most population concentrations. Historically the sea has provided only a small percentage of the world food supply. Increased fish consumption was viewed as a way to meet the needs of a rapidly growing global population.

In fact, during the past half-century per capita consumption of fish has doubled worldwide, and tripled in developing countries (Figure 10.8.4). Still, fish accounts for only 6 percent of all protein consumed by humans, though a rapidly increasing source in developing countries if not in developed ones (Figure 10.8.5).

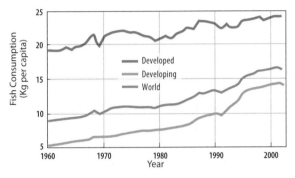

▲ 10.8.4 **FISH CONSUMPTION PER CAPITA**

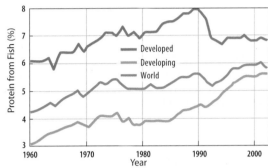

▲ 10.8.5 **PERCENT PROTEIN FROM FISH**

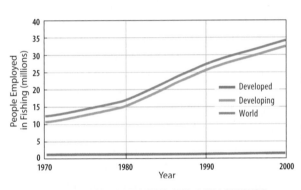

▲ 10.8.6 **EMPLOYMENT IN FISHING AND AQUACULTURE**

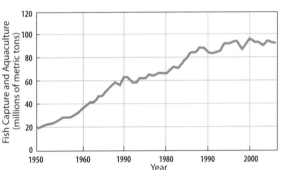

▲ 10.8.7 **WORLD FISH CAPTURE AND AQUACULTURE**

▼ 10.8.8 **AQUACULTURE**
Japan.

OVERFISHING

Worldwide, 35 million people are employed in fishing and agriculture, nearly all in developing countries (Figure 10.8.6). Production of fish is increasing worldwide (Figure 10.8.7). The growth results entirely from expansion of aquaculture (Figure 10.8.8). The capture of wild fish in the oceans and lakes has stagnated since the 1990s despite population growth and increased demand to consume fish.

The population of some fish species in the oceans and lakes has declined because of **overfishing**, which is capturing fish faster than they can reproduce. Overfishing has been particularly acute in the North Atlantic and Pacific oceans. Overfishing has reduced the population of tuna and swordfish by 90 percent in the past half-century, for example. The United Nations estimates that one-quarter of fish stocks have been overfished and one-half fully exploited, leaving only one-fourth underfished.

10.9 Subsistence Agriculture and Population Growth

▶ **Four strategies can increase food supply in developing countries.**

▶ **Increasing productivity and finding new sources are most promising.**

Two issues discussed in earlier chapters influence the challenges faced by subsistence farmers. First, because of rapid population growth in developing countries (discussed in Chapter 2), subsistence farmers must feed an increasing number of people. Second, because of adopting the international trade approach to development (discussed in Chapter 9), subsistence farmers must grow food for export instead of for direct consumption. Four strategies have been identified to increase food supply.

EXPAND AGRICULTURAL LAND

Historically, world food production increased primarily by expanding the amount of land devoted to agriculture. When the world's population increased more rapidly during the Industrial Revolution beginning in the eighteenth century, pioneers could migrate to sparsely inhabited territory and cultivate the land.

New land might appear to be available, because only 11 percent of the world's land area is currently used for agriculture. But excessive or inadequate water makes expansion difficult. The expansion of agricultural land has been much slower than the increase of the human population for several decades (Figure 10.9.1).

▼ 10.9.1
AGRICULTURAL LAND AND POPULATION GROWTH

[Line graph showing Agricultural Land (billion hectares, left axis) and Population (billion people, right axis) from 1960 to 2010. Both axes range 3.0 to 7.0. Agricultural Land rises gently from about 4.5 to 5.0; Population rises steeply from about 3.1 to 6.7.]

Legend: Agricultural Land, Population
X-axis: Year (1960, 1970, 1980, 1990, 2000, 2010)

▼ 10.9.2 **INTERNATIONAL RICE RESEARCH INSTITUTE, HOME OF THE "GREEN REVOLUTION"**

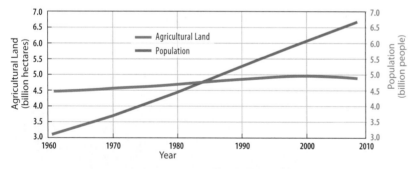

IR64680-81-2-2-1-3
BASAL   24 DAT   PI
45   45   45
Kg N./ha

INCREASE AGRICULTURAL PRODUCTIVITY

New agricultural practices have permitted farmers worldwide to achieve much greater yields from the same amount of land. The invention and rapid diffusion of more productive agricultural techniques during the 1960s and 1970s is called the **green revolution**.

Scientists began experiments during the 1950s to develop a higher-yield form of wheat. A decade later, the International Rice Research Institute created a "miracle" rice seed (Figure 10.9.2). The Rockefeller and Ford foundations sponsored many of the studies, and the program's director, Dr. Norman Borlaug, won the Nobel Peace Prize in 1970. More recently, scientists have developed new high-yield maize (corn). Scientists have continued to create higher-yield hybrids that are adapted to environmental conditions in specific regions.

The green revolution was largely responsible for preventing a food crisis in developing countries during the 1970s and 1980s. The new miracle seeds were diffused rapidly around the world. India's wheat production, for example, more than doubled in 5 years. After importing 10 million tons of wheat annually in the mid-1960s, India by 1971 had a surplus of several million tons.

Will these scientific breakthroughs continue in the twenty-first century? To take full advantage of the new "miracle seeds," farmers must use more fertilizer and machinery, both of which depend on increasingly expensive fossil fuels. To maintain the green revolution, governments in developing countries must allocate scarce funds to subsidize the cost of seeds, fertilizers, and machinery.

IMPROVED FOOD SOURCES

Improved food sources could come from:

- Higher protein cereal grains. People in developing countries depend on grains that lack certain proteins. Hybrids with higher protein content could achieve better nutrition without changing food-consumption habits.

- Palatability of rarely consumed foods. Some foods are rarely consumed because of taboos, religious values, and social customs. In developed countries, consumers avoid consuming recognizable soybean products like tofu and sprouts, but could be induced to eat soybeans shaped like burgers and franks (Figure 10.9.3).

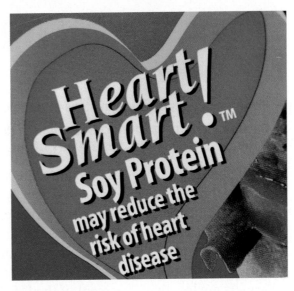

◄ 10.9.3 **SOY PRODUCT**

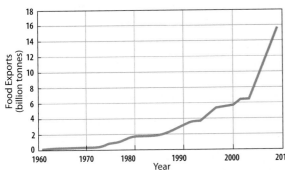

▲ 10.9.4 **WORLD FOOD EXPORTS**

EXPAND EXPORTS

Trade in food has increased rapidly, especially since 2000 (Figure 10.9.4). The three top export grains are wheat, maize (corn), and rice (Figure 10.9.5). Argentina, Brazil, the Netherlands and the United States are the four leading net exporters of agricultural products (Figure 10.9.6). Japan, China, Russia, and the United Kingdom are the leading net importers.

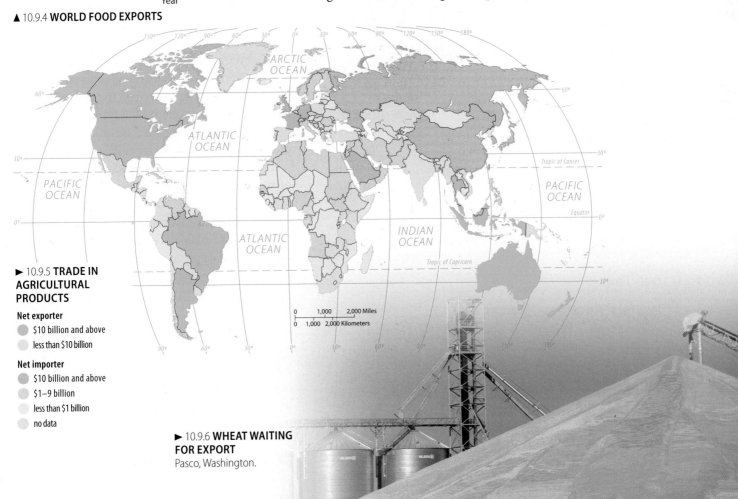

► 10.9.5 **TRADE IN AGRICULTURAL PRODUCTS**

Net exporter
- $10 billion and above
- less than $10 billion

Net importer
- $10 billion and above
- $1–9 billion
- less than $1 billion
- no data

► 10.9.6 **WHEAT WAITING FOR EXPORT**
Pasco, Washington.

10.10 Commercial Agriculture and Market Forces

▶ **Farming is part of agribusiness in developed countries.**

▶ **Because of overproduction, farmers in developed countries may receive government subsidies to reduce output.**

The system of commercial farming found in developed countries is called **agribusiness**, because the family farm is not an isolated activity but is integrated into a large food-production industry. Agribusiness encompasses such diverse enterprises as tractor manufacturing, fertilizer production, and seed distribution. This type of farming responds to market forces rather than to feeding the farmer. Geographers use the von Thünen model to help explain the importance of proximity to market in the choice of crops on commercial farms (Figure 10.10.1).

Farmers are less than 2 percent of the U.S. labor force, but around 20 percent of U.S. labor works in food production and services related to agribusiness—food processing, packaging, storing, distributing, and retailing. Although most farms are owned by individual families, many other aspects of agribusiness are controlled by large corporations.

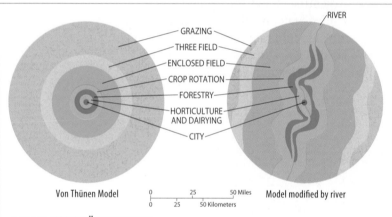

GRAZING
THREE FIELD
ENCLOSED FIELD
CROP ROTATION
FORESTRY
HORTICULTURE AND DAIRYING
CITY
RIVER

Von Thünen Model

0   25   50 Miles
0   25   50 Kilometers

Model modified by river

▲ 10.10.1 **VON THÜNEN MODEL**
Johann Heinrich von Thünen, a farmer in northern Germany, proposed a model to explain the importance of proximity to market in the choice of crops on commercial farms. The von Thünen model was first proposed in 1826 in a book titled *The Isolated State*. According to the model, which was later modified by geographers, a commercial farmer initially considers which crops to cultivate and which animals to raise based on market location. Von Thünen based his general model of the spatial arrangement of different crops on his experiences as owner of a large estate in northern Germany. He found that specific crops were grown in different rings around the cities in the area.

PRODUCTIVITY CHALLENGES

The experience of dairy farming in the United States demonstrates the growth in productivity (Figure 10.10.2). The number of dairy cows has declined since 1960 but production has increased, because yield per cow has tripled (Figure 10.10.3).

Commercial farmers suffer from low incomes because they are capable of producing much more food than is demanded by consumers in developed countries. Although the food supply has increased in developed countries, demand has remained constant, because of low population growth and market saturation (Figure 10.10.4).

A surplus of food can be produced because of widespread adoption of efficient agricultural practices. New seeds, fertilizers, pesticides, mechanical equipment, and management practices have enabled farmers to obtain greatly increased yields per area of land.

▼ 10.10.2 **DAIRY COWS**
California.

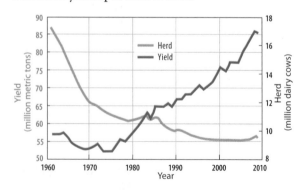

▲ 10.10.3 **U.S. DAIRY PRODUCTIVITY**

What situation factors influence industrial location?

What site factors influence industrial location?

SCAN TO ACCESS
U.S. LABOR
STATISTICS

11.1 The Industrial Revolution

▶ **The Industrial Revolution transformed how goods are produced for society.**
▶ **The United Kingdom was home to key events in the Industrial Revolution.**

The modern concept of industry—meaning the manufacturing of goods in a factory—originated in northern England and southern Scotland during the second half of the eighteenth century. From there, industry diffused in the nineteenth century to Europe and to North America and in the twentieth century to other regions.

ORIGINS OF THE INDUSTRIAL REVOLUTION

The **Industrial Revolution** was a series of improvements in industrial technology that transformed the process of manufacturing goods. Prior to the Industrial Revolution, industry was geographically dispersed across the landscape. People made household tools and agricultural equipment in their own homes or obtained them in the local village. Home-based manufacturing was known as the **cottage industry** system (Figure 11.1.1).

The term *Industrial Revolution* is somewhat misleading:

- The transformation was far more than industrial, and it did not happen overnight.
- The Industrial Revolution resulted in new social, economic, and political inventions, not just industrial ones.

▲▼ 11.1.1 **TRANSFORMATION OF AN INDUSTRY**
(top) In the early nineteenth century, the textile industry was a cottage industry based on people spinning and weaving by hand in their homes. (bottom) By the middle of the century, the industry had become based in factories and mills. In this interior view of a cotton mill in 1835 girls and women tend carding, drawing, and roving machinery.

- The changes involved a gradual diffusion of new ideas and techniques over decades, rather than an instantaneous revolution.

Nonetheless, the term is commonly used to define the process that began in the United Kingdom in the late 1700s.

The invention most important to the development of factories was the steam engine, patented in 1769 by James Watt, a maker of mathematical instruments in Glasgow, Scotland (Figure 11.1.2). Watt built the first useful steam engine, which could pump water far more efficiently than the watermills then in common use, let alone human or animal power. The large supply of steam power available from James Watt's steam engines induced firms to concentrate all steps in a manufacturing process in one building attached to a single power source.

▲ 11.1.2 **JAMES WATT'S STEAM ENGINE**
Steam injected in a cylinder (left side of engine) pushes a piston attached to a crankshaft that drives machinery (right side of engine).

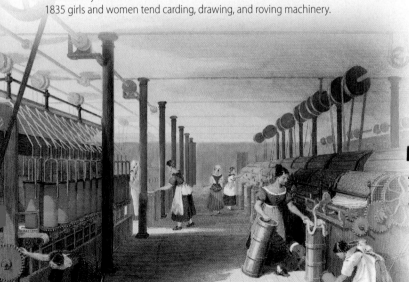

TRANSFORMATION OF KEY INDUSTRIES

Industries impacted by the Industrial Revolution included:

- **Coal:** The source of energy to operate the ovens and the steam engines. Wood, the main energy source prior to the Industrial Revolution, was becoming scarce in England because it was in heavy demand for construction of ships, buildings, and furniture, as well as for heat. Manufacturers turned to coal, which was then plentiful in England.

- **Iron:** The first industry to benefit from Watt's steam engine. The usefulness of iron had been known for centuries, but it was difficult to produce because ovens had to be constantly heated, something the steam engine could do (Figure 11.1.3).

- **Transportation:** Critical for diffusing the Industrial Revolution. First canals and then railroads enabled factories to attract large numbers of workers, bring in bulky raw materials such as iron ore and coal, and ship finished goods to consumers (Figure 11.1.4).

- **Textiles:** Transformed from a dispersed cottage industry to a concentrated factory system during the late eighteenth century, as illustrated in Figure 11.1.1. In 1768, Richard Arkwright, a barber and wigmaker in Preston, England, invented machines to untangle cotton prior to spinning. Too large to fit inside a cottage, spinning frames were placed inside factories near sources of rapidly flowing water, which supplied the power. Because the buildings resembled large watermills, they were known as mills.

- **Chemicals:** An industry created to bleach and dye cloth. In 1746, John Roebuck and Samuel Garbett established a factory to bleach cotton with sulfuric acid obtained from burning coal. When combined with various metals, sulfuric acid produced another acid called vitriol, useful for dying clothing.

- **Food processing:** Essential to feed the factory workers no longer living on farms. In 1810, French confectioner Nicholas Appert started canning food in glass bottles sterilized in boiling water.

▲ 11.1.3 **IRON ORE SMELTING**
Coalbrookdale by Night, an 1801 painting by Philip James de Loutherbourg, depicts the Coalbrookdale Company's iron ore smelter in Ironbridge, England. The painting is in London's Science Museum.

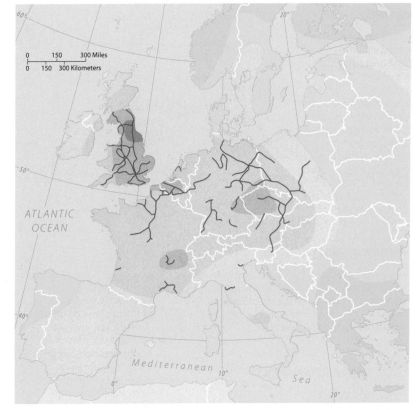

▲ 11.1.4 **DIFFUSION OF RAILROADS**
Europe's political problems retarded the diffusion of the railroad. Cooperation among small neighboring states was essential to build an efficient rail network and to raise money for constructing and operating the system. Because such cooperation could not be attained, railroads in some parts of Europe were delayed 50 years after their debut in the United Kingdom.

First railway opened by
1826 1856
1836 1876
1846 After 1876

 Rail lines constructed by 1848

11.2 Distribution of Industry

▶ **Three-fourths of the world's manufacturing is clustered in three regions.**
▶ **The major industrial regions are divided into subareas.**

Industry is concentrated in three of the nine world regions discussed in Chapter 9: Europe (Figures 11.2.1 and 11.2.2), East Asia (Figure 11.2.3), and North America (Figure 11.2.4). Each of the three regions accounts for roughly one-fourth of the world's total industrial output. Outside these three regions the leading industrial producers are Brazil and India.

▼ 11.2.1 **EUROPE'S INDUSTRIAL AREAS**
Europe was the first region to industrialize during the nineteenth century. Numerous industrial centers developed in Europe as countries competed with each other for supremacy.

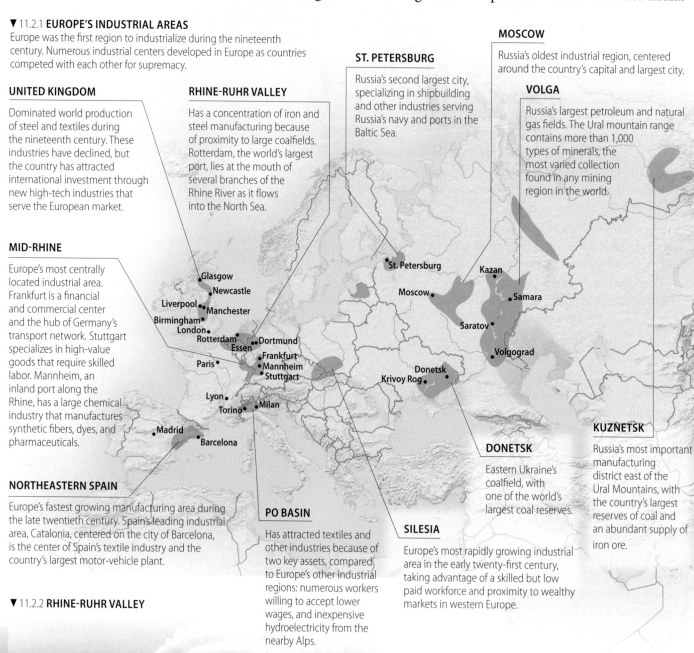

MOSCOW
Russia's oldest industrial region, centered around the country's capital and largest city.

ST. PETERSBURG
Russia's second largest city, specializing in shipbuilding and other industries serving Russia's navy and ports in the Baltic Sea.

UNITED KINGDOM
Dominated world production of steel and textiles during the nineteenth century. These industries have declined, but the country has attracted international investment through new high-tech industries that serve the European market.

RHINE-RUHR VALLEY
Has a concentration of iron and steel manufacturing because of proximity to large coalfields. Rotterdam, the world's largest port, lies at the mouth of several branches of the Rhine River as it flows into the North Sea.

VOLGA
Russia's largest petroleum and natural gas fields. The Ural mountain range contains more than 1,000 types of minerals, the most varied collection found in any mining region in the world.

MID-RHINE
Europe's most centrally located industrial area. Frankfurt is a financial and commercial center and the hub of Germany's transport network. Stuttgart specializes in high-value goods that require skilled labor. Mannheim, an inland port along the Rhine, has a large chemical industry that manufactures synthetic fibers, dyes, and pharmaceuticals.

NORTHEASTERN SPAIN
Europe's fastest growing manufacturing area during the late twentieth century. Spain's leading industrial area, Catalonia, centered on the city of Barcelona, is the center of Spain's textile industry and the country's largest motor-vehicle plant.

PO BASIN
Has attracted textiles and other industries because of two key assets, compared to Europe's other industrial regions: numerous workers willing to accept lower wages, and inexpensive hydroelectricity from the nearby Alps.

SILESIA
Europe's most rapidly growing industrial area in the early twenty-first century, taking advantage of a skilled but low paid workforce and proximity to wealthy markets in western Europe.

DONETSK
Eastern Ukraine's coalfield, with one of the world's largest coal reserves.

KUZNETSK
Russia's most important manufacturing district east of the Ural Mountains, with the country's largest reserves of coal and an abundant supply of iron ore.

Map labels: Glasgow, Newcastle, Liverpool, Manchester, Birmingham, London, Rotterdam, Dortmund, Essen, Frankfurt, Mannheim, Paris, Stuttgart, Lyon, Torino, Milan, Madrid, Barcelona, St. Petersburg, Moscow, Kazan, Samara, Saratov, Volgograd, Krivoy Rog, Donetsk

▼ 11.2.2 **RHINE-RUHR VALLEY**

▼ 11.2.3 EAST ASIA'S INDUSTRIAL AREAS

East Asia became an important industrial region in the second half of the twentieth century, beginning with Japan. In the twenty-first century, China has emerged as the world's leading manufacturing country by most measures.

CHINA

The world's largest supply of low-cost labor and the world's largest market for many consumer products. Manufacturers cluster in three areas along the east coast: near Guangdong and Hong Kong, the Yangtze River valley between Shanghai and Wuhan, and along the Gulf of Bo Hai from Tianjin and Beijing to Shenyang.

Tianjin, Beijing & Shenyang
Shenyang
Beijing
Tianjin
Bo Hai
NORTH KOREA
SOUTH KOREA

CHINA
Nanjing
Shanghai
Wuhan
Yangtze River Valley

Guangdong Province & Hong Kong
Hong Kong

Nagasaki

JAPAN
Tokyo-Yokohama
Nagoya   Tokyo
Kyoto   Yokohama
Kobe   Osaka
Osaka-Kobe-Kyoto

JAPAN

Became an industrial power in the 1950s and 1960s, initially by producing goods that could be sold in large quantity at cut-rate prices to consumers in other countries. Manufacturing is concentrated in the central region between Tokyo and Nagasaki.

▼ 11.2.4 NORTH AMERICA'S INDUSTRIAL AREAS

Industry arrived a bit later in North America than in Europe, but it grew much faster in the nineteenth century. North America's manufacturing was traditionally highly concentrated in northeastern United States and southeastern Canada. In recent years, manufacturing has relocated to the South, lured by lower wages and legislation that has made it difficult for unions to organize factory workers.

SOUTHEASTERN ONTARIO

Canada's most important industrial area, central to the Canadian and U.S. markets and near the Great Lakes and Niagara Falls.

MOHAWK VALLEY

A linear industrial belt in upper New York State, taking advantage of inexpensive electricity generated at nearby Niagara Falls.

NEW ENGLAND

A cotton textile center in the early nineteenth century. Cotton was imported from southern states and finished cotton products were shipped to Europe.

WESTERN GREAT LAKES

Centered on Chicago, the hub of the nation's transportation network, now the center of steel production.

CANADA
NORTH AMERICA
Milwaukee   Detroit   Buffalo   Boston
Chicago   Pittsburgh   New York City
Philadelphia
Baltimore

MIDDLE ATLANTIC

The largest U.S. market, so the region attracts industries that need proximity to a large number of consumers and depends on foreign trade through one of this region's large ports.

SOUTHERN CALIFORNIA

Now the country's largest area of clothing and textile production, the second-largest furniture producer, and a major food-processing center.

San Francisco
Los Angeles

PITTSBURGH-LAKE ERIE

The leading steel-producing area in the nineteenth century because of proximity to Appalachian coal and iron ore.

MEXICO

11.3 Situation Factors in Locating Industry

▶ **A manufacturer typically faces two geographical costs: situation and site.**
▶ **Situation factors involve transporting materials to and from a factory.**

Geographers explain why one location may prove more profitable for a factory than others. Situation factors are discussed in the next four sections, and site factors later in the chapter. **Situation factors** involve transporting materials to and from a factory. A firm seeks a location that minimizes the cost of transporting inputs to the factory and finished goods to consumers.

PROXIMITY TO INPUTS

Every industry uses some inputs and sells to customers. The farther something is transported, the higher the cost, so a manufacturer tries to locate its factory as close as possible to both buyers and sellers.

- If inputs are more expensive to transport than products, the optimal location for a factory is near the source of inputs.

- If the cost of transporting the product to customers exceeds the cost of transporting inputs, then the optimal plant location is as close as possible to the customer.

Every manufacturer uses some inputs. These may be resources from the physical environment (minerals, wood, or animals), or they may be parts or materials made by other companies. An industry in which the inputs weigh more than the final products is a **bulk-reducing industry**. To minimize transport costs, a bulk-reducing industry locates near the source of its inputs. An example is copper production (Figure 11.3.1). Copper ore is very heavy when mined, so mills that concentrate the copper by removing less valuable rock are located close to the mine.

▲ 11.3.1 **BULK-REDUCING INDUSTRY: COPPER**
Use Google Earth to explore Mount Isa, Australia's leading copper production center, which includes mining and smelting.
Fly to: *Parkside, Mount Isa, Australia.*

What is the large crater-like feature in the image?
Locate the plume of smoke just west of Parkside and drag the street view icon to the main road west of the Parkside label.

What type of structure is producing the smoke? Why is this structure located close to the other feature?

JUST-IN-TIME DELIVERY

Proximity to market has long been important for many types of manufacturers, as discussed earlier in this chapter. The factor has become even more important in recent years because of the rise of just-in-time delivery.

As the name implies, **just-in-time** is shipment of parts and materials to arrive at a factory moments before they are needed. Just-in-time delivery is especially important for delivery of inputs, such as parts and raw materials, to manufacturers of fabricated products, such as cars and computers.

Under just-in-time, parts and materials arrive at a factory frequently, in many cases daily if not hourly. Suppliers of the parts and materials are told a few days in advance how much will be needed over the next week or two, and first thing each morning exactly what will be needed at precisely what time that day.

Just-in-time delivery reduces the money that a manufacturer must tie up in wasteful inventory (Figure 11.6.4). The percentage of the U.S. economy tied up in inventory has been cut in half during the past quarter-century. Manufacturers also save money through just-in-time delivery by reducing the size of the factory, because space does not have to be wasted on piling up a mountain of inventory.

To meet a tight timetable, a supplier of parts and materials must locate factories near its customers. If only an hour or two notice is given, a supplier has no choice but to locate a factory within 50 miles or so of the customer.

▲ 11.6.4 **ELIMINATING INVENTORY**
Leading computer manufacturers have cut costs in part through eliminating the need to store inventory in warehouses. These computers are being built in China only after the buyer has placed the order.

Just-in-time delivery sometimes merely shifts the burden of maintaining inventory to suppliers. Walmart, for example, holds low inventories but tells its suppliers to hold high inventories "just in case" a sudden surge in demand requires restocking on short notice.

JUST-IN-TIME DISRUPTIONS

Just-in-time delivery means that producers have less inventory to cushion against disruptions in the arrival of needed parts. Three kinds of disruptions can result from reliance on just-in-delivery:

- **Labor unrest.** A strike at one supplier plant can shut down the entire production within a couple of days. Also disrupting deliveries could be a strike in the logistics industry, such as truckers or dockworkers.

- **Traffic.** Deliveries may be delayed when traffic is slowed by accident, construction, or unusually heavy volume. Trucks and trains are both subject to these types of delays , especially crossing international borders (Figure 11.6.5).

- **Natural hazards.** Poor weather conditions can afflict deliveries anywhere in the world. Blizzards and floods can close highways and rail lines. The 2011 earthquake and tsunami in Japan put many factories and transportation lines out of service for months. Carmakers around the world had to curtail production because key parts had been made at the damaged factories.

► 11.6.5 **DELIVERY DISRUPTIONS**
These vehicles on a highway in Ontario are backed up trying to cross the border into Michigan.

11.7 Site Factors in Industry

▶ **Site factors result from the unique characteristics of a location.**

▶ **The three main site factors are labor, land, and capital.**

Site factors are industrial location factors related to the costs of factors of production inside the plant, notably labor, land, and capital.

LABOR

A **labor-intensive** industry is one in which wages and other compensation paid to employees constitute a high percentage of expenses. Labor costs an average of 11 percent of overall manufacturing costs in the United States, so a labor-intensive industry would have a much higher percentage than that (Figure 11.7.1).

The average annual wage paid to male workers exceeds $30,000 or $15 per hour in most developed countries, compared to less than $5,000 or $2.50 per hour in most developing countries (Figure 11.7.2). Health care, retirement pensions, and other benefits add substantially to the wage compensation in developed countries, but not in developing countries.

For some manufacturers—but not all—the difference between paying workers $2.50 and $15 per hour is critical. For example, most

▲ 11.7.1 **LABOR**
Chinese workers in a packaging factory. Around the world, approximately 150 million people are employed in manufacturing, according to the UN International Labor Organization (ILO). China has around 20 percent of the world's manufacturing workers and the United States around 10 percent.

▲ 11.7.2 **EARNED ANNUAL INCOME (MALES)**

- ● $30,000 and above
- ● $10,000 – $29,999
- ● $5,000 – $9,999
- ○ below $5,000
- ○ no data

of the cost of an iPhone is in the parts (made mostly in Japan, Germany, and South Korea) and the gross profit to Apple (based in the United States). One step in the production process is labor intensive—snapping all the parts together at an assembly plant—and this step is done in China with relative low-wage workers (Figure 11.7.3).

▶ 11.7.3 **COST STRUCTURE OF AN iPHONE**
The cost of manufacturing an iPhone is substantially less than the price that consumers pay.

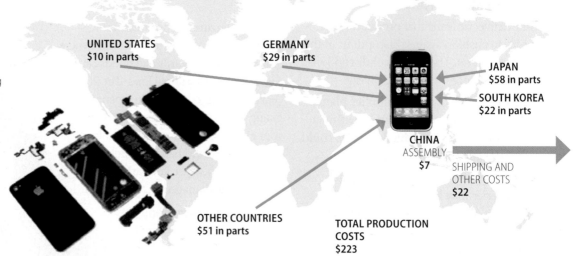

UNITED STATES
$10 in parts

GERMANY
$29 in parts

JAPAN
$58 in parts

SOUTH KOREA
$22 in parts

CHINA
ASSEMBLY
$7

SHIPPING AND OTHER COSTS
$22

OTHER COUNTRIES
$51 in parts

TOTAL PRODUCTION COSTS
$223

LAND

In the early years of the Industrial Revolution, multistory factories were constructed in the heart of the city. Now, they are more likely to be built in suburban or rural areas, in part to provide enough space for one-story buildings.

Raw materials are typically delivered at one end and moved through the factory on conveyors or forklift trucks. Products are assembled in logical order and shipped out at the other end.

Locations on the urban periphery are also attractive for factories to facilitate delivery of inputs and shipment of products. In the past, when most material moved in and out of a factory by rail, a central location was attractive because rail lines converged there.

With trucks now responsible for transporting most inputs and products, proximity to major highways is more important for a factory. Especially attractive is the proximity to the junction of a long-distance route and the beltway or ring road that encircles most cities. Factories cluster in industrial parks located near suburban highway junctions (Figure 11.7.4).

Also, land is much cheaper in suburban or rural locations than near the center city. A hectare (or an acre) of land in the United States may cost only a few thousand dollars in a rural area, tens of thousands in a suburban location, and hundreds of thousands near the center.

▲ 11.7.4 **LAND**
Factory outside Vic, Spain.

CAPITAL

Manufacturers typically borrow capital—the funds to establish new factories or expand existing ones. One important factor in the clustering in California's Silicon Valley of high-tech industries has been availability of capital. One-fourth of all capital in the United States is spent on new industries in the Silicon Valley (Figure 11.7.5).

Banks in Silicon Valley have long been willing to provide money for new software and communications firms even though lenders elsewhere have hesitated. High-tech industries have been risky propositions—roughly two-thirds of them fail—but Silicon Valley financial institutions have continued to lend money to engineers with good ideas so that they can buy the software, communications, and networks they need to get started.

The ability to borrow money has become a critical factor in the distribution of industry in developing countries. Financial institutions in many developing countries are short of funds, so new industries must seek loans from banks in developed countries. But enterprises may not get loans if they are located in a country that is perceived to have an unstable political system, a high debt level, or ill-advised economic policies.

▼ 11.7.5 **CAPITAL**
San Jose, California, in Silicon Valley.

11.8 Textile and Apparel Production

▶ Textile and apparel production is a prominent example of a labor-intensive industry.
▶ Textile and apparel production generally requires less skilled, low-wage workers.

Production of apparel and textiles, which are woven fabrics, is a prominent example of an industry that generally requires less-skilled, low-cost workers. The textile and apparel industry accounts for 6 percent of the dollar value of world manufacturing but a much higher 14 percent of world manufacturing employment, an indicator that it is a labor-intensive industry. The percentage of the world's women employed in this type of manufacturing is even higher.

Textile and apparel production involves three principal steps. All are labor-intensive compared to other industries, but the importance of labor varies somewhat among them. As a result, their global distributions are not identical, because the three steps are not equally labor-intensive.

SPINNING OF FIBERS TO MAKE YARN

Fibers can be spun from natural or synthetic elements (Figure 11.8.1). Cotton is the principal natural fiber—three-fourths of the total—followed by wool. Historically, natural fibers were the sole source, but today synthetics account for three-fourths and natural fibers only one-fourth of world thread production. Because it is a labor-intensive industry, spinning is done primarily in low-wage countries, primarily China (Figure 11.8.2).

▲ 11.8.1 **COTTON YARN PRODUCTION, CHINA**
China produces two-thirds of the world's cotton yarn.

▶ 11.8.2 **COTTON YARN PRODUCTION**

Metric tons
- 100,000 and above
- 10,000 – 99,999
- 1,000 – 9,999
- below 1,000
- no data

ARCTIC OCEAN

ATLANTIC OCEAN

PACIFIC OCEAN

ATLANTIC OCEAN

INDIAN OCEAN

PACIFIC OCEAN

Tropic of Cancer

Equator

Tropic of Capricorn

0   1,000   2,000 Miles
0   1,000   2,000 Kilometers

WEAVING OR KNITTING YARN INTO FABRIC

Labor constitutes an even higher percentage of total production cost for weaving than for the spinning and assembly steps. China alone accounts for nearly 60 percent of the world's woven cotton fabric production, and India another 30 percent (Figure 11.8.3) Fabric has been woven or laced together by hand for thousands of years on a loom, which is a frame on which two sets of threads are placed at right angles to each other. Even on today's mechanized looms, a loom has one set of threads, called a warp, which is strung lengthwise. A second set of threads, called a weft, is carried in a shuttle that moves over and under the warp (Figure 11.8.4).

CUTTING AND SEWING OF FABRIC FOR ASSEMBLING INTO CLOTHING AND OTHER PRODUCTS

Textiles are assembled into four main types of products: garments, carpets, home products such as bed linens and curtains, and industrial uses such as headliners inside motor vehicles. Developed countries play a larger role in assembly than in spinning and weaving because most of the consumers of assembled products are located there. For example, two-thirds of the women's blouses sold worldwide in a year are sewn in developed countries (Figure 11.8.5). However, the percentage of clothing produced in developing countries has been increasing.

▲ 11.8.3 **WOVEN COTTON FABRIC PRODUCTION**

Square meters
- 20 billion and above
- 1 billion – 3 billion
- 0.1 billion – 1.0 billion
- below 0.1 billion
- no data

◄ 11.8.4 **WOVEN COTTON FABRIC PRODUCTION, CHINA**

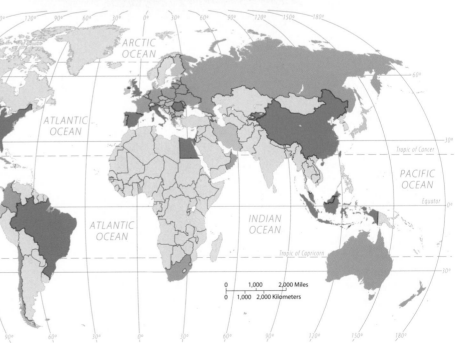

► 11.8.5 **WOMEN'S BLOUSE PRODUCTION**
- 10 million and above
- 1 million – 9 million
- 100,000 – 999,999
- below 100,000
- no data

11.9 Emerging Industrial Regions

► Manufacturing is growing in locations not traditionally considered as industrial centers.

► The four BRIC countries are expected to be increasingly important industrial centers.

Industry is on the move around the world. Site factors, especially labor costs, have stimulated industrial growth in new regions, both internationally and within developed regions. Situation factors, especially proximity to growing markets, have also played a role in the emergence of new industrial regions.

INTERREGIONAL SHIFTS IN THE UNITED STATES

Manufacturing jobs have been shifting within the United States from the North and East to the South and West (Figure 11.9.1). Between 1950 and 2009, the North and East lost 6 million manufacturing jobs and the South and West gained 2 million.

The principal site factor for many manufacturers was labor-related: enactment of **right-to-work** laws by a number of states, especially in the South. A right-to-work law requires a factory to maintain a so-called "open shop" and prohibits a "closed shop."

• In a "closed shop," a company and a union agree that everyone must join the union to work in the factory.

• In an "open shop," a union and a company may not negotiate a contract that requires workers to join a union as a condition of employment.

By enacting right-to-work laws, southern states made it much more difficult for unions to organize factory workers, collect dues, and bargain with employers from a position of strength. As a result, the percentage of workers who are members of a union is much lower in the South than elsewhere in the United States. Car plants, steel, textiles, tobacco products, and furniture industries have dispersed through smaller communities in the South, many in search of a labor force willing to work for less pay than in the North and to forgo joining a union (Figure 11.9.2).

▼ 11.9.1 CHANGING U.S. MANUFACTURING, 1950 AND 2010

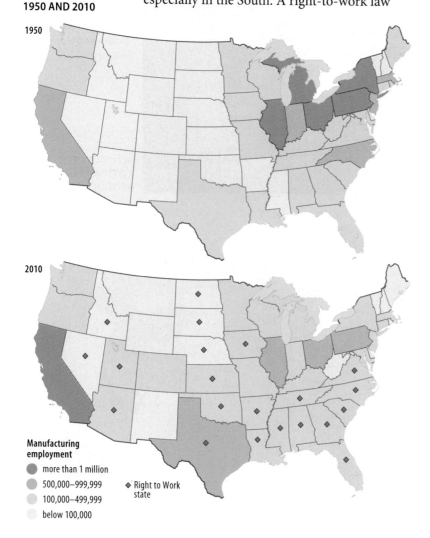

1950

2010

Manufacturing employment

- more than 1 million
- 500,000–999,999
- 100,000–499,999
- below 100,000

◆ Right to Work state

▲ 11.9.2 VEHICLE ASSEMBLY PLANTS OPEN AND CLOSED BETWEEN 1980 AND 2010

- Plant opened
- Plant closed

CALIFORNIA

INDUSTRY IN MEXICO

Manufacturing has been increasing in Mexico. The North Atlantic Free Trade Agreement (NAFTA), effective 1994, eliminated most barriers to moving goods between Mexico and the United States.

Because it is the nearest low-wage country to the United States, Mexico attracts labor-intensive industries that also need proximity to the U.S. market. Although the average wage is higher in Mexico than in most developing countries (refer to Figure 11.7.2), the cost of shipping from Mexico to the United States is lower than from other developing countries.

Mexico City, the country's largest market, is the center for industrial production for domestic consumption (Figure 11.9.3). Other factories have located in Mexico's far north to be as close as possible to the United States.

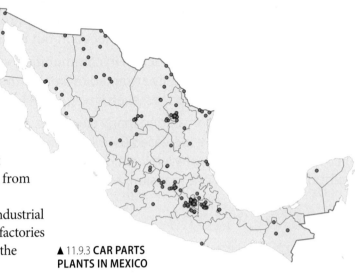

▲ 11.9.3 **CAR PARTS PLANTS IN MEXICO**

EMERGING INDUSTRIAL POWERS: THE "BRIC" COUNTRIES

Much of the world's future growth in manufacturing is expected to locate outside the principal industrial regions described earlier in section 11.2. The financial analysis firm Goldman Sachs has coined the acronym BRIC to indicate the countries it expects to dominate global manufacturing during the twenty-first century. BRIC is an acronym for four countries—Brazil, Russia, India, and China (Figure 11.9.4). They are also known as the newly emerging economies.

The four BRIC countries together currently control one-fourth of the world's land and two-fifths of the world's population, but the four combined account for only one-sixth of world GDP. All four countries have made changes to their economies in recent years, embracing international trade with varying degrees of enthusiasm. By mid-twenty-first century, the four BRIC countries, plus the United States and Mexico, are expected to have the world's six largest economies.

The four BRIC countries have different advantages for industrial location. Russia and Brazil, (Figure 11.9.5) currently classified by the United Nations as having high levels of development, are especially rich in inputs critical for industry. China and India, classified as having medium levels of development, have the two largest labor forces and potential markets for consumer goods.

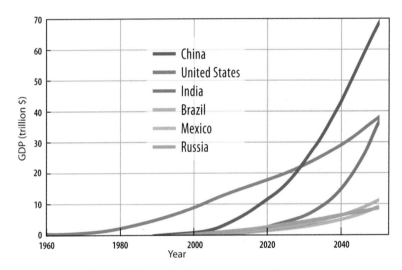

▲ 11.9.4 **GDP HISTORY AND FORECAST FOR BRIC COUNTRIES, UNITED STATES, AND MEXICO**

► 11.9.5 **COMPUTER MANUFACTURING PLANT, CURITIBA, BRAZIL**

Three recent changes in the structure of manufacturing have geographic consequences:

• Factories have become more productive through introduction of new machinery and processes. A factory may continue to operate at the same location but require fewer workers to produce the same output.

• Companies are locating production in communities where workers are willing to adopt more flexible work rules. Firms are especially attracted to smaller towns where low levels of union membership reduce vulnerability to work stoppages, even if wages are kept low and layoffs become necessary.

• By spreading production among many countries, or among many communities within one country, large corporations have increased their bargaining power with local governments and labor forces. Production can be allocated to locations where the local government is especially helpful and generous in subsidizing the costs of expansion, and the local residents are especially eager to work in the plant.

Key Questions

Where is industry clustered?

▶ The Industrial Revolution originated in the United Kingdom and diffused to Europe and North America in the twentieth century.

▶ World industry is highly clustered in three regions—Europe, North America, and East Asia.

What situation factors influence industrial location?

▶ A company tries to identify the optimal location for a factory through analyzing situation and site factors.

▶ Situation factors involve the cost of transporting both inputs into the factory and products from the factory to consumers.

▶ Steel and motor vehicle industries have traditionally located factories primarily because of situation factors.

What site factors influence industrial location?

▶ Three site factors—land, labor, and capital—control the cost of doing business at a location.

▶ Production of textiles and apparel has traditionally been located primarily because of site factors.

▶ New industrial regions are emerging because of their increased importance for site and situation factors.

Thinking Geographically

The North American Free Trade Agreement (NAFTA) among Canada, Mexico, and the United States was implemented in 1994.

1. **What have been the benefits and the drawbacks to Canada, Mexico, and the United States as a result of NAFTA?**

To induce Hyundai to open a plant in 2010 in West Point, Georgia, to assemble its Kia models, the state spent $36 million to buy the site and donate it to Hyundai, $61 million to build infrastructure such as roads and rail lines, $73 million to train the workers, and $90 million in tax benefits (Figure 11.CR.1).

2. **Why would the state of Georgia spend $260 million to get the Kia factory? Did Georgia overpay?**

Manufacturing is more dispersed than in the past, both within and among countries.

3. **What are the principal manufacturers in your community or area? How have they been affected by increasing global competition?**

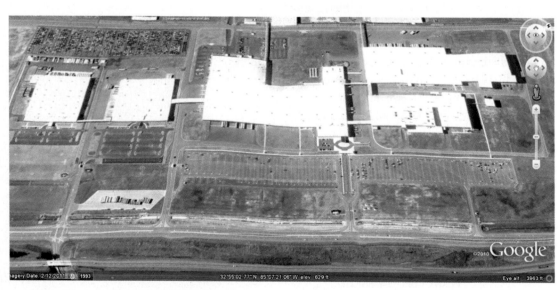

▶ 11.CR.1 **KIA ASSEMBLY PLANT, WEST POINT, GEORGIA**

Interactive Mapping

SITUATION FACTORS AND RUSSIAN INDUSTRY

Russia's principal industrial location asset is proximity to inputs.

Launch MapMaster Russian Domain in Mastering**GEOGRAPHY**™

Select *Economic* then *Industrial Regions.*

Next select *Economic* then *Major Natural resources* then *Coal and Iron only* (deslect others).

Coal and iron are the two principal inputs into steel production. Which of these inputs is close to Russia's principal industrial areas, and which must be transported relatively far?

Deselect coal and iron and instead select other natural resources to see which are near the industrial areas and which have to be transported.

What overall pattern of industrial location do you see?

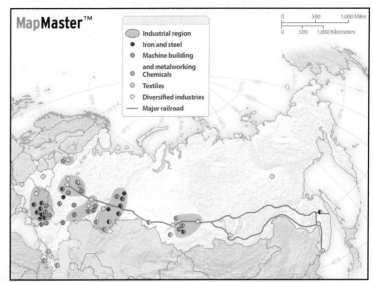

MapMaster™

- Industrial region
- Iron and steel
- Machine building and metalworking
- Chemicals
- Textiles
- Diversified industries
- Major railroad

Explore

SAN ANTONIO, TEXAS

Use Google Earth to explore the changing industrial landscape in the United States.

Fly to: *1 Lone Star Pass, San Antonio, Texas, USA.*

Zoom out until the entire factory complex is visible.

Click: *Show Historical Imagery.*

Move the time line to: *9/27/2002.*

What has changed from 9/27/2002 until today?

Move the time line forward from 9/27/2002; at what date can a change in the land first be seen?

Zoom out again until the city of San Antonio is visible.

What are some advantages of 1 Lone Star Pass as an industrial location?

1 Lone Star Pass, San Antonio, TX 78221

Key Terms

Break-of-bulk point
A location where transfer is possible from one mode of transportation to another.

Bulk-gaining industry
An industry in which the final product weighs more or comprises a greater volume than the inputs.

Bulk-reducing industry
An industry in which the final product weighs less or comprises a lower volume than the inputs.

Cottage industry
Manufacturing based in homes rather than in a factory, commonly found prior to the Industrial Revolution.

Industrial Revolution
A series of improvements in industrial technology that transformed the process of manufacturing goods.

Just-in-time delivery
Shipment of parts and materials to arrive at a factory moments before they are needed.

Labor-intensive industry
An industry for which labor costs comprise a high percentage of total expenses.

Right-to-work state
A U.S. state that has passed a law preventing a union and company from negotiating a contract that requires workers to join a union as a condition of employment.

Site factors
Location factors related to the costs of factors of production inside the plant, such as land, labor, and capital.

Situation factors
Location factors related to the transportation of materials into and from a factory.

On the Internet

Statistics on employment in manufacturing, as well as other sectors of the U.S. economy, are at the U.S. Department of Labor's Bureau of Labor Statistics website, at **www.bls.gov**, or scan the QR on the first page of this chapter.

▶ LOOKING AHEAD

Most of the growth in jobs in the United States and in the world are in the service (or tertiary) sector, and most service jobs are located in urban settlements.

12 Services and Settlements

Flying across the United States on a clear night, you look down on the lights of settlements, large and small. You see small clusters of lights from villages and towns, and large, brightly lit metropolitan areas. Geographers apply economic geography concepts to explain regularities in the pattern of settlements.

The regular pattern of settlements in the United States and other developed countries reflects where services are provided. Three-fourths of the workers in developed countries are employed in the service sector of the economy. These services are provided in settlements.

The regular distribution of settlements observed over developed countries is not seen in developing countries. Geographers explain that the pattern in developing countries results from having much lower percentages of workers in services.

Where are consumer services distributed?

12.1 **Types of Services**

12.2 **Central Place Theory**

12.3 **Hierarchy of Consumer Services**

12.4 **Market Area Analysis**

INTERNET CAFE,
THAILAND

THE COSTS OF SPRAWL

When private developers select new housing sites, they seek cheap land that can easily be prepared for construction—land often not contiguous to the existing built-up area. Land is not transformed immediately from farms to housing developments. Instead, developers buy farms for future construction of houses by individual builders. Developers frequently reject land adjacent to built-up areas in favor of sites outside the urbanized area, depending on the price and physical attributes of the alternatives. The periphery of U.S. cities therefore looks like Swiss cheese, with pockets of development and gaps of open space (Figure 13.9.2).

Urban sprawl has some undesirable traits:

• Roads and utilities must be extended to connect isolated new developments to nearby built-up areas.

• Motorists must drive longer distances and consume more fuel.

• Agricultural land is lost to new developments, and other sites lie fallow while speculators await the most profitable time to build homes on them.

• Local governments typically spend more on services for these new developments than they collect in additional taxes.

SEGREGATION

The modern residential suburb is segregated in two ways.

• Housing in a given suburb is usually built for people of a single social class, with others excluded by virtue of the cost, size, or location of the housing. Segregation by race and ethnicity also persists in many suburbs (see Sections 7.2 and 7.3).

• Residents are separated from commercial and manufacturing activities that are confined to compact, distinct areas.

SUBURBAN RETAILING

Suburban residential growth has fostered change in the distribution of consumer services. Historically, urban residents bought food and other daily necessities at small neighborhood shops in the midst of housing areas and shopped in the CBD for other products. CBD sales have stagnated because suburban residents won't make the long journey there.

Instead, retailing has been increasingly concentrated in planned suburban shopping malls, auto-friendly strip malls, and big-box stores, surrounded by generous parking lots (Figure 13.9.3). These nodes of consumer services are called **edge cities.** Edge cities originated as suburban residences for people who worked in the central city, and then shopping malls were built to be near the residents. Edge cities now also serve as nodes of business services (Figure 13.9.4).

A shopping center is built by a developer, who buys the land, builds the structures, and leases space to individual merchants. The key to a successful large shopping center is the inclusion of one or more anchors. Most consumers go to a center to shop at an anchor and, while there, patronize the smaller shops. The anchors may be a supermarket and discount store in a smaller center or several department stores in a larger center.

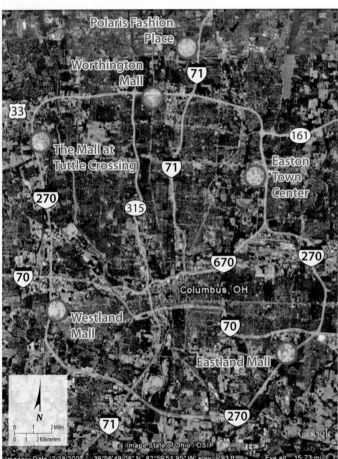

▲ 13.9.3 **SHOPPING MALLS NEAR COLUMBUS, OHIO**

▼ 13.9.4 **EDGE CITY**
Easton Town Center, outside Columbus, Ohio.

13.10 Urban Transportation

▶ **Most trips in the U.S. are by private motor vehicle.**
▶ **Public transportation has made a modest comeback in some cities.**

People do not travel aimlessly; their trips have a precise point of origin, destination, and purpose. More than half of all trips are work related. Shopping or other personal business and social journeys each account for approximately one-fourth of all trips. Sprawl makes people more dependent on motor vehicles for access to work, shopping, and social activities.

DEVELOPMENT OF URBAN TRANSPORTATION

Historically, people lived close together in cities because they had to be within walking distance of shops and places of employment. The invention of the railroad in the nineteenth century enabled people to live in suburbs and work in the central city. Cities then built street railways (called trolleys, streetcars, or trams) and underground railways (subways) to accommodate commuters. Rail and trolley lines restricted suburban development to narrow strips within walking distance of the stations.

▲ 13.10.1 **HIGHWAYS IN SAN FRANCISO**

MOTOR VEHICLES

▼ 13.10.2 **CONGESTION CHARGING IN LONDON**

Until the twentieth century, the growth of suburbs was constrained by poor transportation. Motor vehicles have permitted large-scale development of suburbs at greater distances from the center, in the gaps between the rail lines. More than 95 percent of all trips within U.S. cities are made by car.

The U.S. government has encouraged the use of cars and trucks by paying 90 percent of the cost of limited-access high-speed interstate highways, which crisscross 74,000 kilometers (46,000 miles) across the country (Figure 13.10.1). The use of motor vehicles is also supported by policies that keep the price of fuel below the level found in most other countries.

The motor vehicle is an important user of land in the city. An average city allocates about one-fourth of its land to roads and parking lots (refer to Figure 13.1.1). Valuable land is devoted to parking cars and trucks, although expensive underground and multistory parking structures can reduce the amount of ground-level space needed. Freeways cut a wide path through the heart of cities, and elaborate interchanges consume even more space.

Motor vehicles have costs beyond their purchase and operation: delays imposed on others, increased need for highway maintenance, construction of new highways, and pollution. The average American loses 36 hours per year sitting in traffic jams and wastes 55 gallons of gasoline.

Technological improvements may help traffic flow. Computers mounted on the dashboards alert drivers to traffic jams and suggest alternate routes. On freeways, vehicle speed and separation from other vehicles can be controlled automatically rather than by the driver.

Motorists can be charged for using congested roads or pay high tolls to drive on uncongested roads (Figure 13.10.2). The inevitable diffusion of such technology in the twenty-first century will reflect the continuing preference of most people to use private motor vehicles.

PUBLIC TRANSIT

In larger cities, public transportation is better suited than motor vehicles to moving large numbers of people, because each traveler takes up far less space. Public transportation is cheaper, less polluting, and more energy efficient than the automobile. It also is particularly suited to rapidly bringing a large number of people into a small area. Despite the obvious advantages of public transportation for commuting, only 5 percent of trips in U.S. cities are by public transit. Outside of big cities, public transportation is extremely rare or nonexistent.

Public transportation has been expanded in some U.S. cities to help reduce air pollution and conserve energy. New subway lines and existing systems expanded in a number of cities (Figure 13.10.3). The federal government has permitted Boston, New York, and other cities to use funds originally allocated for interstate highways to modernize rapid transit service instead. The trolley—now known by the more elegant term of light-rail transit—is making a modest comeback in North America (Figure 13.10.4). California, the state that most symbolizes the automobile-oriented American culture, is the leader in construction of new light-rail transit lines, as well as retention of historic ones (Figure 13.10.5).

Despite modest recent successes, most public transportation systems are caught in a vicious circle, because fares do not cover operating costs. As patronage declines and expenses rise, the fares are increased, which drives away passengers and leads to service reduction and still higher fares.

▲ 13.10.3 **PUBLIC TRANSIT OPTIONS IN SAN FRANCISCO: BART SUBWAY**

◄ 13.10.4 **PUBLIC TRANSIT OPTIONS IN SAN FRANCISCO: MUNI LIGHT RAIL**

▼ 13.10.5 **PUBLIC TRANSIT OPTIONS IN SAN FRANCISCO: CABLE CARS**

What is the future for cities? As shown in this chapter, contradictory trends are at work simultaneously. Why does one inner-city neighborhood become a slum and another an upper-class district? Why does one city attract new shoppers and visitors while another languishes?

The suburban lifestyle as exemplified by the detached single-family house with surrounding yard attracts most people. Yet inner-city residents may rarely venture out to suburbs. Lacking a motor vehicle, they have no access to most suburban locations. Lacking money, they do not shop in suburban malls or attend sporting events at suburban arenas. The spatial segregation of inner-city residents and suburbanites lies at the heart of the stark contrasts so immediately observed in any urban area. Several U.S. states have taken strong steps in the past few years to curb sprawl, reduce traffic congestion, and reverse inner-city decline. The goal is to produce a pattern of compact and contiguous development, while protecting rural land for agriculture, recreation, and wildlife.

Key Questions

Where are people distributed within urban areas?

▶ The Central Business District (CBD) contains a large share of a city's business and public services.

▶ The concentric zone, sector, and multiple nuclei models describe where different types of people live within urban areas.

▶ The three models together foster understanding that people live in different rings, sectors, and nodes depending on their stage in life, social status, and ethnicity.

How are urban areas expanding?

▶ Urban areas have expanded beyond the legal boundaries of cities to encompass urbanized areas and metropolitan areas that are functionally tied to the cities.

▶ With suburban growth, most metropolitan areas have been fragmented into a large number of local governments.

What challenges do cities face?

▶ Most Americans now live in suburbs that surround cities.

▶ Low-income inner-city residents face a variety of economic, social, and physical challenges.

▶ Tying together sprawling American urban areas is dependency on motor vehicles.

Thinking Geographically

Draw a sketch of your community or neighborhood. In accordance with Kevin Lynch's *The Image of the City*, place five types of information on the map—districts (homogeneous areas), edges (boundaries that separate districts), paths (lines of communication), nodes (central points of interaction), and landmarks (prominent objects on the landscape).

1. How clear an image does your community have for you?

Jane Jacobs wrote in *Death and Life of Great American Cities* that an attractive urban environment is one that is animated with an intermingling of a variety of people and activities, such as found in many New York City neighborhoods (Figure 13.CR.1).

2. What are the attractions and drawbacks to living in such environments?

Officials of rapidly growing cities in developing countries discourage the building of houses that do not meet international standards for sanitation and construction. Also discouraged are private individuals offering transportation in vehicles that lack decent tires, brakes, and other safety features. Yet the residents prefer substandard housing to no housing, and they prefer unsafe transportation to no transportation.

3. What would be the advantages and problems for a city if health and safety standards for housing, transportation, and other services were relaxed?

▶ 13.CR.1 **NEW YORK'S GREENWICH VILLAGE**

Interactive Mapping

OVERLAPPING METROPOLITAN AREAS IN NORTH AMERICA

Overlapping metropolitan areas are emerging in North America in addition to Megalopolis.

Open MapMaster North America in Mastering**GEOGRAPHY**™

Select: *Political* then *Cities*

Select: *Population* then *Population Density.*

Where in North America other than Megalopolis do there appear to be overlapping metropolitan areas?

Explore

CHICAGO, ILLINOIS

Use Google Earth to explore Chicago's changing lakefront.

Fly to: *Soldier Field, Chicago.*

Drag to enter Street View on top of Soldier Field.

Exit Ground level View.

Rotate compass so that North is at the top.

Zoom out and move the image until the lakefront and green peninsula (Northerly Island Park) are visible to the east and buildings to the west.

Click Historical Imagery and slide date back to 4/23/2000

1. **What changes have occurred to the east of Soldier Field, along the lakefront?**
2. **What changes have occurred to the west of Soldier Field?**
3. **How do you think the development of this node of the city will influence people's activities?**

Key Terms

Annexation
Legally adding land area to a city in the United States.

Census tract
An area delineated by the U.S. Bureau of the Census for which statistics are published; in urbanized areas, census tracts are often delineated to correspond roughly to neighborhoods.

Central business district (CBD)
The area of a city where consumer, business, and public services are clustered.

City
An urban settlement that has been legally incorporated into an independent, self-governing unit.

Combined statistical area (CSA)
In the United States, two or more contiguous core based statistical areas tied together by commuting patterns.

Concentric zone model
A model of the internal structure of cities in which social groups are spatially arranged in a series of rings.

Core based statistical area (CBSA)
In the United States, a term referring to either a metropolitan statistical area or a micropolitan statistical area.

Edge city
A large node of office and retail activities on the edge of an urban area.

Gentrification
A process of converting an urban neighborhood from a predominantly low-income renter-occupied area to a predominantly middle-class owner-occupied area.

Metropolitan statistical area (MSA)
In the United States, a central city of at least 50,000 population, the county within which the city is located, and adjacent counties meeting one of several tests indicating a functional connection to the central city.

Micropolitan statistical area (μSA)
In the United States, an urban area of between 10,000 and 50,000 inhabitants, the county in which it is found, and adjacent counties tied to the city

Multiple nuclei model
A model of the internal structure of cities in which social groups are arranged around a collection of nodes of activities.

Peripheral model
A model of North American urban areas consisting of an inner city surrounded by large suburban residential and business areas tied together by a beltway or ring road.

Primary census statistical area (PCSA)
In the United States, all of the combined statistical areas plus all of the remaining metropolitan statistical areas and micropolitan statistical areas.

Sector model
A model of the internal structure of cities in which social groups are arranged around a series of sectors, or wedges, radiating out from the central business district (CBD).

Social area analysis
Statistical analysis used to identify where people of similar living standards, ethnic background, and lifestyle live within an urban area.

Sprawl
Development of new housing sites at relatively low density and at locations that are not contiguous to the existing built-up area.

Squatter settlement
An area within a city in a developing country in which people illegally establish residences on land they do not own or rent and erect homemade structures.

Urbanized area
In the United States, a central city plus its contiguous built-up suburbs.

On the Internet

Social Explorer provides access to census data at all scales, including urban, at **www.socialexplorer.com**, or scan the QR at the beginning of the chapter. An interactive map enables users to choose the area of interest from among hundreds of census variables.

▶ **LOOKING AHEAD**

Our journey ends with an examination of the use, misuse, and reuse of resources.

14 Resource Issues

People transform Earth's land, water, and air for their benefit. But human actions in recent years have gone far beyond actions of the past. With less than one-fourth of the world's population, developed countries consume most of the world's energy and generate most of its pollutants. Meanwhile in developing countries, 2 billion people live without clean water or sewers, and 1 billion live in cities with unsafe sulfur dioxide levels.

Geographers study the troubled relationship between human actions and the physical environment in which we live. From the perspective of human geographers, Earth offers a large menu of resources available for people to use. The problem is that most resources are limited, and Earth has a tremendous number of consumers. Geographers observe two major misuses of resources:

- We deplete scarce resources, especially petroleum, natural gas, and coal, for energy production.

- We destroy resources through pollution of air, water, and soil.

These two misuses are the basic themes of this chapter.

NUCLEAR POWER STATION, HAMM-UENTROP, GERMANY